DIAGRID STRUCTURES

SYSTEMS / CONNECTIONS / DETAILS

TERRI MEYER BOAKE

DIAGRID STRUCTURES

SYSTEMS / CONNECTIONS / DETAILS

BIRKHÄUSER / BASEL

The preparation of this book was kindly supported
by the World Steel Association.

worldsteel
ASSOCIATION

GRAPHIC DESIGN, LAYOUT AND TYPOGRAPHY
Rein Steger, Pau Segarra
Proxi, Barcelona

EDITOR FOR THE PUBLISHER
Andreas Müller, Berlin

LIBRARY OF CONGRESS
CATALOGING-IN-PUBLICATION DATA
A CIP catalog record for this book has been
applied to at the Library of Congress.

BIBLIOGRAPHIC INFORMATION PUBLISHED
BY THE DEUTSCHE NATIONALBIBLIOTHEK
The Deutsche Nationalbibliothek lists this publication in the Deutsche Nationalbibliografie; detailed bibliographic data are available in the Internet at http://dnb.dnb.de.

P.O. Box 44, 4009 Basel, Switzerland
Part of De Gruyter

Printed on acid-free paper produced
from chlorine-free pulp. TCF ∞

Printed in Germany

ISBN 978-3-03821-564-6
9 8 7 6 5 4 3 2 1

www.birkhauser.com

CONTENTS

FOREWORD

The World Steel Association (worldsteel) is proud to be the exclusive sponsor of *Diagrid Structures: Systems/Connections/Details*, showcasing the cutting-edge use of steel in architectural design. Diagrid structures are located in North America, Europe, the Middle East, Asia and Australasia, truly representing a significant worldwide use of innovative steel. Construction activities absorb large amounts of steel every year because of the strength, formability and versatility of steel in many different applications. Indeed, we believe steel to be at the base of global economic activity.

Steel is a key driver of the world's economy. The industry directly employs more than two million people worldwide, with a further two million contractors and four million people in supporting industries. It is the strength of steel that has allowed for the creation of vibrant cities and buildings of all shapes and sizes that provide inspirational places to live, work and play.

The tensile capabilities of steel surpass all other common construction materials. Steel has supported the quest for building height and bridge span. Diagrid structures fully exploit the capabilities of steel and in doing so create a new, unique type of structural system that enables a limitless range of architectural expression.

Sustainable steel is at the core of a green economy. Steel, created as long ago as 150 years, can be recycled and reused in new products and applications. The amount of energy required to produce a tonne of steel has been reduced by 50% in the past 30 years, making it a good choice for building structures. 97% of steel by-products can be reused. As we look to the future life of our buildings through "design for disassembly", steel will allow for the immediate reuse of structural elements. What cannot be reused as is can be recycled into a closed-loop, cradle-to-cradle system. Steel does not waste.

Steel is everywhere in our lives, from food cans, appliances and automobiles to the buildings we work and live in. The housing and construction sector is the largest consumer of steel today, using around 50% of the world's steel production.

Steel is safe, innovative and progressive. The industry is committed to the safety and health of its people and is committed to the goal of an injury-free workplace. Nowhere is this more important than in the construction industry, where ironworkers are handling multiple tonnes of steel and working at great heights. One of the advantages of diagrid construction is its use of prefabricated elements. Their design uses the most advanced digital design and fabrication software. Prefabricated diagrid elements can often be sub-assembled at the construction site, thereby limiting the work to be done at height.

We believe that this book highlights collaboration opportunities between steel and construction.

Dr. Edwin Basson
Director General, World Steel Association

This isometric drawing of the skeleton of The Leadenhall Building in London, England, also as shown on the book cover, demonstrates the essential role of structural steel in the creation of a diagrid building system.

PREFACE

Diagrids have emerged as one of the most innovative and adaptable approaches to creating building structures in this millennium. As a construction system it is highly dependent on the capabilities of steel in terms of design, (pre)fabrication and erection processes. Effective collaboration between the architect, engineer and steel fabricator/erector is also critical to the success of a diagrid project.

This text expands upon research initiated in *Understanding Steel Design: An Architectural Design Manual*, published in 2011. As the majority of diagrid buildings have been completed since 2002, and are increasing in number, this seemed a suitable topic for a first volume of further exploration in cutting edge developments in steel design. Of the many buildings visited during the research I conducted over the last dozen years to write this book, each is unique in its form and adaptation of the diagrid. Yet all have certain elements in common. The identification of these common elements and approaches to design have provided the basis for the structure of this book.

The design of a diagrid structure takes a very particular approach that modifies the more standard approach to orthogonal steel structures. This includes the transformation of steel connections into nodal points - *nodes* - that are often prefabricated to best accommodate non-standard and often-changing geometries. An overall modularity - *module* - is used to determine the placement of the nodes, which in turn are connected by straight members. The angularity necessarily influences *façade design*. The ability of the diagrid to assume all lateral loading allows for different considerations in the design and materiality of the *service core*. The interdependence of these components creates a decision-making process that is quite different from more standard structural types.

The book is therefore divided into three parts: first to examine the factors that led to the creation of the early diagrid buildings (Chapters 2 and 3); second to comparatively analyze a significant number of recent diagrid buildings to establish some precedents that can be used to assist with the design of diagrid buildings (Chapters 4 to 10); third to take a more detailed, focused look at several recent projects (Chapter 11).

Many of the photos used in the creation of the text were taken by me during my travels. Where I was unable to visit a project at a critical phase, or for important construction and fabrication photographs and drawings, images have been sourced from the engineers, fabricators and architects involved. These are specifically credited at the back of this text. Many thanks for your contribution of drawings and information. They lent to a wonderfully complex and rich research project.

This book would not have been possible without the generous sponsorship of the World Steel Association.

ACKNOWLEDGEMENTS

Particular thanks to Arup's London office for providing me a space to conduct research and access to their extensive library and image bank.

The far-reaching content would not have been possible to elaborate without the assistance and generosity of many people. Significant contributors include: ArcelorMittal, Ateliers Jean Nouvel, Brookfield Multiplex Australasia, Peter Chipchase (Arup), Council on Tall Buildings and Urban Habitat, D+H Steel, Damian Eley (Arup), Elizabeth C. English, Neb Erakovic (Yolles), Foster + Partners, Alistair Fussell (Steel Construction New Zealand), Bryan Hamilton (Aedas), William Hare Structural Engineers, Vincent Hui, Leslie E. Robertson Associates, London Legacy, Kyoung Sun Moon, Miroslav Munka, Ahmad Rahimian (WSP Cantor Seinuk), Rogers Stirk Harbour + Partners, RUR Architecture, David Ryan (Australian Steel Institute), Jeff Schofield (ADNEC), Skidmore, Owings & Merrill, TFP Architects, Walters Inc., Warren and Mahoney Architects and Chris Wilkinson (Wilkinson Eyre Architects).

Additional gratitude to Sylvie Boulanger, Walter Koppelaar and Tim Verhey for inspiration and information. Also thanks to my family who supports my continuous wanderings in search of state-of-the-art architecture.

Thank you to the designers of diagrid buildings for creating such innovative and inspirational architecture.

TIMELINE

This timeline looks at the series of projects that are addressed in this book in the context of the development of the diagrid, from the early work of Vladimir Shukhov, through the diagonalized core typology, to the present. This is by no means a complete list of all of the diagrid buildings constructed to date. The application of the diagonalized core from its introduction in the 1960s to the present is very extensive. The sampling is global and intended to provide a thorough overview of the development of the system.

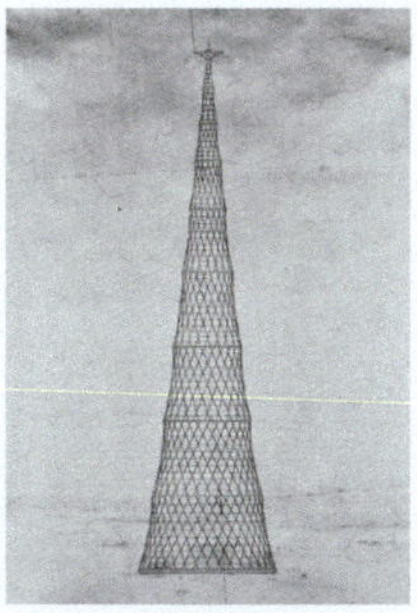

1896–1919

SHUKHOV TOWERS
(VARIOUS) RUSSIA

Building Height: up to 350m/1,148ft
Diagrid Type: exposed lattice
Designer: Vladimir Shukhov

1963

IBM BUILDING (UNITED IRONWORKERS)
PITTSBURGH, PA, USA

Building Height: 13 floors
Diagrid Type: concealed diagrid
Architect: Curtis and Davis Architects
Engineer: Leslie E. Robertson Associates

1969

JOHN HANCOCK CENTER
CHICAGO, IL, USA

Building Height: 344m/1,128ft/100 floors
Diagrid Type: diagonalized core
Architect: SOM
Engineer: SOM

1990

BANK OF CHINA TOWER
HONG KONG, CHINA

Building Height: 367m/1,205ft/72 floors
Diagrid Type: diagonalized core
Architect: I.M. Pei
Engineer: Leslie E. Robertson Associates

1996

PUERTA DE EUROPA
MADRID, SPAIN

Building Height: 114m/374ft/26 floors
Diagrid Type: diagonalized core
Architect: Johnson/Burgee Architects
Engineer: Leslie E. Robertson Associates

2002

LONDON CITY HALL
LONDON, ENGLAND

Building Height: 10 floors
Diagrid Type: AESS diagrid to support glazing
Architect: Foster + Partners
Engineer: Arup

2004

SWISS RE (30 ST. MARY AXE)
LONDON, ENGLAND

Building Height: 180m/590ft/40 floors
Diagrid Type: concealed diagrid
Architect: Foster + Partners
Engineer: Arup

2006

HEARST MAGAZINE TOWER
NEW YORK CITY, NY, USA

Building Height: 182m/597ft/46 floors
Diagrid Type: concealed diagrid
Architect: Foster + Partners
Engineer: WSP Cantor Seinuk

2006

ROYAL ONTARIO MUSEUM ADDITION
TORONTO, ON, CANADA

Building Height: 6 floors
Diagrid Type: concealed diagrid
Architect: Libeskind w/ Bregman and Hamman
Engineer: Arup

2008

CANTON TOWER
GUANGZHOU, CHINA

Building Height: 600m/1,969ft
Diagrid Type: sightseeing tower, external AESS diagrid
Architect: Mark Hemel/Barbara Kuit/IBA
Engineer: Arup

2008

TORNADO TOWER
DOHA, QATAR

Building Height: 195m/640ft/51 floors
Diagrid Type: partially exposed diagrid
Architect: CICO Consulting Architects and Engineers, SAIT
Engineer: Stroh and Ernst AG

2011

CAPITAL GATE
ABU DHABI, UAE

Building Height: 165m/540ft/36 floors
Diagrid Type: AESS diagrid
Architect: RMJM
Engineer: RMJM

2011

KK100
SHENZHEN, CHINA

Building Height: 442m/1,499ft/100 floors
Diagrid Type: diagonalized core
Architect: TFP Architects
Engineer: Arup

2012

AL BAHAR TOWERS
ABU DHABI, UAE

Building Height: 145m/476ft/29 floors
Diagrid Type: honeycomb
Architect: Aedas
Engineer: Arup

2012

CCTV
BEIJING, CHINA

Building Height: 234m/768ft/54 floors
Diagrid Type: concealed and AESS diagrid
Architect: Rem Koolhaas (OMA)
Engineer: Arup

2012

ONE SHELLEY STREET
SYDNEY, AUSTRALIA

Building Height: 11 floors
Diagrid Type: concealed diagrid
Architect: Fitzpatrick + Partners
Engineer: Arup

2014

MANUKAU INSTITUTE OF TECHNOLOGY
AUCKLAND, NEW ZEALAND

Building Height: 5 floors
Diagrid Type: AESS diagrid to support glazing
Architect: Warren and Mahoney Architects
Engineer: Holmes Consulting Group

2010

GUANGZHOU IFC
GUANGZHOU, CHINA

Building Height: 439m/1,439ft/103 floors
Diagrid Type: AESS diagrid
Architect: Wilkinson Eyre Architects
Engineer: Arup

2010

O-14
DUBAI, UAE

Building Height: 106m/347ft/24 floors
Diagrid Type: concrete diagrid variation
Architects: RUR Architecture (Reiser + Umemoto)
Engineer: Ysrael A. Seinuk

2011

ALDAR HEADQUARTERS
ABU DHABI, UAE

Building Height: 110m/361ft/25 floors
Diagrid Type: concealed diagrid
Architect: MZ Architects
Engineer: Arup

2012

ARCELORMITTAL ORBIT TOWER
LONDON, ENGLAND

Building Height: 115m/377ft
Diagrid Type: AESS diagrid
Architect: Anish Kapoor, Cecil Balmond
Engineer: Arup

2012

BOW ENCANA TOWER
CALGARY, AB, CANADA

Building Height: 237m/779ft/57 floors
Diagrid Type: AESS diagrid
Architect: Foster + Partners w/ Zeidler Partnership
Engineer: Yolles

2012

DOHA TOWER
DOHA, QATAR

Building Height: 238m/781ft/46 floors
Architect: Ateliers Jean Nouvel
Diagrid Type: AESS diagrid
Engineer: Terrell Group, China Construction Design International

2014

THE LEADENHALL BUILDING
LONDON, ENGLAND

Building Height: 224m/735ft/50 floors
Diagrid Type: AESS diagrid
Architect: Rogers Stirk Harbour + Partners
Engineer: Arup

2015

LOTTE SUPER TOWER
SEOUL, SOUTH KOREA

Building Height: 555m/1,819ft/112 floors
Diagrid Type: vision – not built
Architect: SOM
Engineer: SOM

2016

ZHONGGUO ZUN TOWER
BEIJING, CHINA

Building Height: 528m/1,732ft/108 floors
Diagrid Type: vision – not built
Architect: TFP Architects
Engineer: Arup

1

A COLLABORATIVE PROCESS

WHAT IS A DIAGRID?

FROM SHUKHOV TO FOSTER

THE IMPORTANCE OF COLLABORATION

THE ROLE OF BUILDING INFORMATION MODELING

WHY CHOOSE A DIAGRID?

DIAGRID DECISIONS, STEP BY STEP

The 2004 completion of the Swiss Re Tower (30 St. Mary Axe) in London, England, designed by Foster + Partners with Arup, marked the beginning of the evolution and application of the modern diagrid building using a perimeter support system.

WHAT IS A DIAGRID?

Over the last 10 years, diagrid structures have proven to be highly adaptable in structuring a wide range of building types, spans and forms. In most applications, diagrids provide structural support to buildings that are non-rectilinear, adapting well to highly angular buildings and curved forms. The diagrid in its purest form is capable of resisting all of the gravity loads and lateral loads on the structure without assistance of a traditional structural core. This permits unique deviations from structural types that are dependent on a core for stability.

The term "diagrid" is a blending of the words "diagonal" and "grid" and refers to a structural system that is single-thickness in nature and gains its structural integrity through the use of triangulation. Diagrid systems can be planar, crystalline or take on multiple curvatures; they often use crystalline forms or curvature to increase their stiffness. Being single-thickness differentiates a diagrid from any three-dimensional triangulated systems such as spaceframes, space trusses or geodesic structures, although it will be shown that some of the developments of diagrid structures have been derived from the details of these three-dimensional systems. The diagrid structural systems that will be explored in this book are used in the support of buildings, predominantly as perimeter systems that are associated with mid to high-rise buildings. Perimeter diagrids normally carry the lateral and gravity loads of the building and are used to support the floor edges.

Diagrid systems are also used as roofs to create large column-free spans. Diagrid systems for this function have been derived from lamella structures, which may be made from a variety of materials but predominantly use wood. The majority of lamella structures, however, are not diagrid structures, as they use a diamond grid and tend not to triangulate. The structural ideas behind the wood lamella contributed to the evolution of the steel lattice grid. The detailing of the steel lattice grid system is significantly different from that of the perimeter structural diagrid for larger buildings. This type of structure was addressed in my previous book *Understanding Steel Design: An Architectural Design Manual* in *Chapter 12: Steel and Glazing Systems*. The design and technical exploration of diagrid structures addressed in this book will build on the introductory material addressed in *Chapter 9: Advanced Framing Systems: Diagrids* of the same book.

FROM SHUKHOV TO FOSTER

The origins of the diagrid structural typology lie at the crossroads of engineering and architecture. Initial explorations by the Russian engineer Vladimir Shukhov were intended to provide a structural system to serve a civic works function that was not necessarily "architecture" in the purest sense of the word. The initial details and member choices were fairly utilitarian and simple. It is significant that Norman Foster has referenced the work of Shukhov as an inspiration for his diagrid explorations. This affirms the role of Shukhov's towers as a precedent for buildings such as the Swiss Re Tower and the Hearst Magazine Tower. It also allows us to examine the changes that were made to the method of detailing and construction as the hyperbolic paraboloid form transitioned from a "hollow" tower to one that needed to support floor loads and was clad. This was a tremendous change in the role of the structure, with significant implications on the design, detailing and construction processes undertaken by Foster and Arup in Swiss Re. The decisions taken in the design of Swiss Re and the Hearst Magazine Tower continue to inform all variations of the diagrid to the present day.

THE IMPORTANCE OF COLLABORATION

Collaborative dialogue is even more critical when using a diagrid than with most other more traditional structural forms. Where Architecturally Exposed Structural Steel (AESS) is used, this necessarily increases the need for communication as the member choice and fabrication details become critical in conveying the architectural aesthetic. For both concealed and AESS diagrid structures, the collaborative dialogue remains crucial, because of the complexity of the structure and also the relative lack of experience that the majority of architects, engineers and fabricators have in the design and construction of a diagrid. There is significant repetition in the engineers involved in the projects included in this book, due to the emerging design challenges and body of knowledge, while code-related issues that govern the seismic design of diagrids remain in the development stage.

During the increased use of AESS over the last two decades, the management, work process and design flow of projects has changed. Where concealed steel has become routine in terms of its details and construction practices, expressed steel requires the input of the fabricator to a much higher degree. Experienced steel fabricators are able to advise the team on issues of detailing, member selection, connections and erection procedures that can save significant time and money on a project. Although diagrids are said to be able to save 20% of the weight of steel used – this varies by project and is not to be taken as an absolute – the engineering and fabrication costs can be significantly higher than in traditional framed and/or concealed structural steel buildings. Even in cases where the steel of the diagrid is not exposed, the design of the nodes and the nature of their on-site connections to the diagrid members will be critical in terms of fabrication costs and ease of assembly. It is interesting that most published articles about highly complex steel buildings fail to mention the steel fabricators and erectors when in fact they have played a critical role.

THE ROLE OF BUILDING INFORMATION MODELING

Technical drawings have changed significantly over the last 30 years. The preference for orthogonal steel-framed buildings can be in part attributed to 20th century limitations in representation and calculation. Most steel-framed buildings could be calculated as determinate structures, resolving most three-dimensional force systems into planar ones. The type of forms presented by Shukhov's hyperbolic paraboloid towers would likely have been beyond the facilities of most engineers of his time and throughout the Modern period.

It is no coincidence that the birth of the contemporary diagrid building type came at a time when computer-assisted drawing was hitting its stride. The development of Building Information Modeling (BIM) has been critical to ensuring the successful design and fabrication of highly complex structures. Geometric complexity requires a very high level of accuracy and makes the assessment of loading on members and connections far more challenging than rectilinear structures that can be reduced to determinate transfers of load. Where computer programs such as Catia allowed architects like Frank Gehry to create irregular curved forms, BIM programs such as Xsteel were created to assist in engineering, detailing and fabricating the steel that supported these unusual forms. The transition of Xsteel to Tekla Structures in 2004 added significantly more functionality and interoperability to the software. Interoperability of the workflow is critical to a successful collaboration among the architect, engineer and steel fabricator. This type of software is under constant development to improve functionality and interoperability.

This technical drawing sheet of Shukhov's Shabolovka Radio Tower, design drawing of 1919, demonstrates the level of detailing and the member choices that were typical of these original diagrid buildings. The construction details emerged from the diagonalized shape as the tower was erected. At the time, erectors were accustomed to looking at the intention of the details and making adjustments as the project progressed.

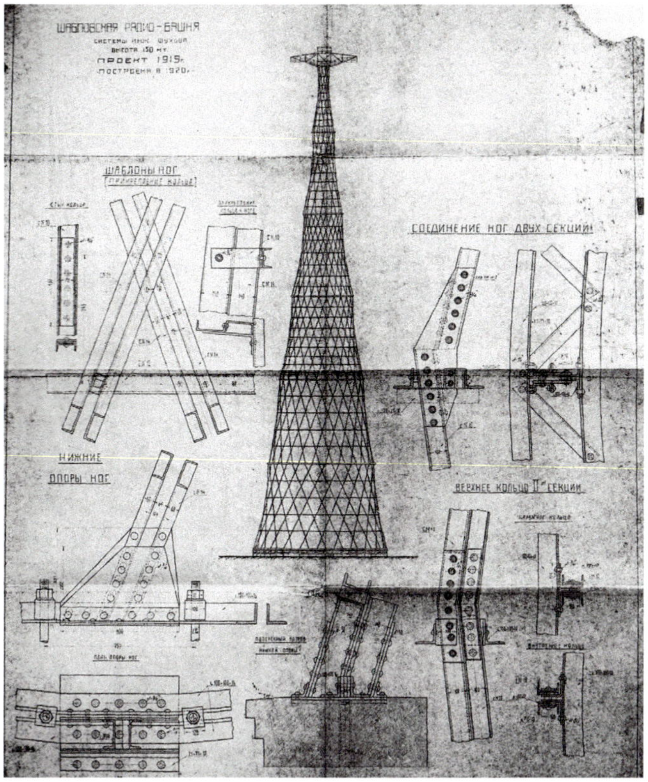

The structural drawings produced using Tekla Structures, Bentley Systems and other software perform a critical role in the design and construction process. These drawings are created by the steel fabricator and used to detail the connections in the project, using structural information provided by the engineer. It is possible to zoom into the 3D model from which the drawing is derived to create or examine each and every connection, permitting the design team to see and share the connection design.

This three-dimensional image of the ground floor framing of the Bow Encana Tower in Calgary, AB, Canada, designed by Foster + Partners, illustrates the type of BIM model that is used by the steel fabricator in the design, detailing and production of shop drawings for a complex diagrid project. The drawing was created by Walters Inc., one of the steel fabricators for this project.

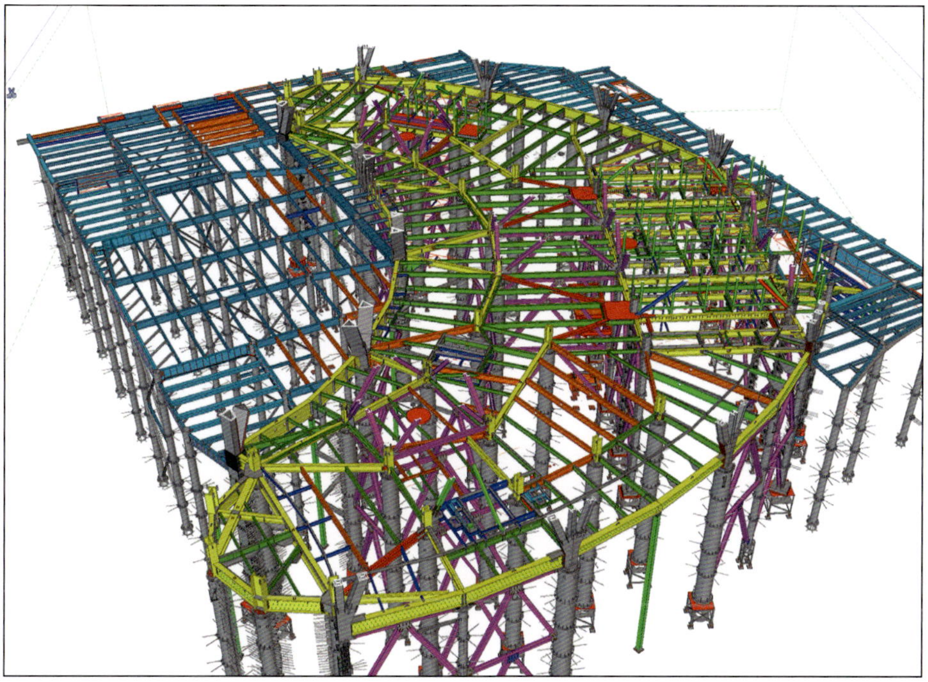

This fabrication drawing for the Royal Ontario Museum in Toronto, ON, Canada, designed by Daniel Libeskind and detailed by Walters Inc., the steel fabricator of the project, allows a glimpse into the level of detail that is part of the specialized BIM steel detailing packages. The digital models incorporate the proper sizes and thicknesses of the members and allow the fabricators to understand the complexity of the structural system in order to properly work out the bolted and welded connections.

This sort of model can be an effective replacement for very expensive physical mock-ups, which are also liable to delay the project delivery. The color format allows the fabricator to clearly identify systems and create drawing sets that define the erection sequence. The shop drawings that are required for the production of each and every member and connection are extracted from this model. The use of this sort of system helps to ensure a high level of accuracy in the fabrication of the elements, which in turn assists in achieving tighter tolerances and better erection.

WHY CHOOSE A DIAGRID?

Diagrids have emerged as an architectural choice in the creation of contemporary buildings. Although there are engineering-based reasons that would suggest the use of a diagrid, discussions with engineers would conclude that architectural design has been the driving motivation. Diagrids are able to adapt to a wide range of non-rectilinear geometric forms, including irregular curves and angles. No other type of framed structure is capable of this task.

There are several functional and economic advantages that underlie the system:

- increased stability due to triangulation
- combination of the gravity and lateral load-bearing systems, potentially providing more efficiency
- provision of alternate load paths (redundancy) in the event of a structural failure
- reduced use of structural materials translating into environmental savings
- reduced weight of the superstructure can translate into a reduced load on the foundations
- ability to provide structural support for a myriad of shapes

From the perspective of the management of the design and construction of the project, there are also aspects in the selection of a diagrid structure that can be very positive:

- → the need for a cooperative team approach between the architect, engineer and fabricator, given the complexity of the design and the level of visual impact of the structure on the building design
- → integration of expertise and specialization from both architects and engineers
- → ability to reduce dependency on the core for achieving lateral stability
- → high level of prefabrication with increased use of shop facilities
- → a higher formal flexibility, given the innate stability of the frame

These procedures and outcomes are not necessarily present in each project. The examination of a wide variety of diagrid applications in this book will look at these on a case-by-case basis.

DIAGRID DECISIONS, STEP BY STEP

The structure of this book is arranged to address a full range of detailed considerations that need to form part of the project discussions.

First-tier decisions:

- → architectural versus structural motivation
- → architectural form
  - → curvilinear versus angular
  - → regular versus irregular
- → height and relative size or scale of the project
- → use of the building
- → fire protection strategy
  - → architecturally expressed steel
  - → system must be protected or concealed

Second-tier decisions:

- → implications of the size of the primary diamond module
- → node design
  - → expressed or exposed steel systems
  - → concealed steel systems
- → core design
  - → no additional lateral load assistance required
  - → core needed to provide additional lateral resistance (tall buildings)
- → issues of constructability
  - → regular versus irregular geometries
  - → staging areas
  - → transportation and access
  - → shop versus site connections
- → façade design
  - → impact of the expression of the diagrid on the façade
  - → rectilinear glazing patterns
  - → triangular glazing patterns
- → exterior diagrids
  - → placement of the diagrid structure outside of the thermal wall
  - → use of diagrid to support a double façade

A wide range of projects will be referenced throughout the sections of the book. Several of these have been presented as Project Profiles where more comprehensive information has been made available through the generosity of the architects, engineers, contractors and fabricators involved in the projects.

2
EVOLUTION OF DIAGRID FRAMING SYSTEMS

Many of the diagrid towers of Vladimir Shukhov, constructed as early as 1896, still stand as the first examples of the steel diagrid type.

Diagonalized grid structures have emerged as one of the most innovative and adaptable approaches to structuring buildings in this millennium. The use of diagrids as a contemporary formal structural language for buildings started in the early 2000s, examples being the London City Hall (2002) and the Swiss Re Tower (2004), also in London, and the Hearst Magazine Tower (2006) in New York City. Foster + Partners was the design architect of all three buildings and Arup responsible for engineering the London projects, with WSP Cantor Seinuk providing engineering services for Hearst. Much like the earlier High Tech work of the Foster office, these projects stood out as quite unique in the way that their structural language permeated all aspects of the design. The particular engineering, fabrication and erection challenges that naturally follow when deviating from fairly standard structural steel technologies subsequently require a higher level of collaboration for these types of projects. Foster's previous innovation and exploration in High Tech architecture had already created a firm with strong collaboration skills.

BIRTH OF THE DIAGRID IN RUSSIAN CONSTRUCTIVISM

In spite of the recent surge in the construction of diagrid buildings, their origins date back close to 100 years. Norman Foster references the work of the Russian Constructivist Vladimir Shukhov (1853–1939) as the precedent for his diagrid tower concepts.[1] Shukhov was one of the most prolific engineers and architects of his time, responsible for the creation of hundreds of civil structures, bridges and long-span trussed roofs.

The creation of a new structural system by the Russian Constructivists, one that was strikingly different from standard framing methods of the past, resulted from a number of converging factors. Industrialization was in full swing by the latter part of the 19th century. The invention of structural steel using the Bessemer method toward the latter part of the 19th century had resulted in a material that was infinitely stronger than any that had preceded it. Steel was capable of being mass-produced in a range of standard profiles and sections, ensuring consistency in the mechanical properties of the product. In particular, steel had highly superior tensile properties in comparison to wrought or cast iron. This enabled experimentation in new kinds of structural forms that were not limited to optimal loading for compression. There was no longer any need to avoid tensile stresses and it was soon discovered that there were benefits in exploiting the tensile capability and visual lightness of this new material.

The invention of the steel skeleton separated the structural function of the frame from the exterior curtain wall, so that by the end of the 1800s steel became the dominant structural material for building tall. Technical developments in reinforced concrete followed steel, but it was not until much later in the 20th century that pumping mechanisms enabled reinforced concrete construction to achieve great heights.

Iron truss structures had come into use early in the 1800s, first in bridge construction and later adapted for large-span roofs. These structures were typically constructed of wrought iron as it performed significantly better in tension than cast iron (although weak in comparison to modern steel). In 1847 the American engineer Squire Whipple authored *A Work on Bridge Building*,[2] which provided analytical methods for establishing the forces in trusses. The inherent rigidity of triangulated structures had been recognized as far back as the third century BCE by the Greeks and was also mentioned in Vitruvius'

Ten Books on Architecture. However, prior to the invention of modern steel most structures could not take advantage of the potential benefit of creating axial tensile forces: the majority of buildings being constructed during the Industrial Revolution adhered to structural types that had been used for centuries. When during the 1800s the field of structural engineering clearly emerged and the roles of the architect and engineer diverged in practice, more complex buildings provided a challenge for engineers, considering the primitive calculation tools of the time. Certainly engineering and physics were applied to structural design in a predictive way, but the underlying mathematics still depended on Euclidian geometry as its basis, ill-suited to addressing non-rectilinear structures.

In 1829 the Russian mathematician Nikolai Ivanovich Lobachevsky published a disproof of Euclid's fifth or parallel postulate using the case of a doubly curved surface, thereby establishing non-Euclidian geometry. Although his new "imaginary geometry" or "pangeometry" was not accepted widely at the time, it did influence those who followed. According to historical research conducted by Elizabeth C. English, there was a direct connection between the mathematical discoveries of Lobachevsky and the structural explorations of Vladimir Shukhov.[3]

First diagrid tower designed by Vladimir Shukhov in 1896.

The "idea of the diagrid" and the first constructed diagrid structure have been credited to Vladimir Shukhov. The design evolved as an efficient and easily constructed tower for carrying a large gravity load at the top – a water tower. The "Shukhov Tower", currently located in Polibino, Russia, and designed in 1896, relies on the use of a diagonal lattice of steel angles, constrained laterally at specific intervals along the height of the tower by steel rings. The overall narrowing of the structure from base to top follows a parabolic curve. Made up of a delicate lattice structure, the tower has five interlocking "hyperboloids" that decrease in size, evoking an inverted telescope.[4] The gently curved form emerges from the slenderness of the steel angles and the restraint provided by the rings. The tower is hollow, requiring little resistance to wind loads, and ascent is achieved via a caged steel ladder. Shukhov is credited with the construction of some 200 towers using this method during his career.

Although the slender steel sections, when formed into the hyperbolic paraboloid shape, did undergo some bending, a diagrid shape emerges, using a combination of straight segments that are joined at their nodal points of intersection. The Shukhov towers tended to use much longer steel sections and overlap them at their crosspoints, rather than using the crosspoints as "nodes" in the fashion of later geodesic or spaceframe structures.

What is very important about this discovery and its application to towers of significant height was that the diagrid form could support both the gravity loads and the lateral loads without requiring additional means. In comparison, the new skyscraper types that were under development at this time used a steel frame to support the gravity loads of the exterior wall and the floors, while the central core provided the stiffness required to resist wind loads. Where additional resistance was required, it was usually the core that carried the bracing.

Another important innovation was Shukhov's extrapolation of the structural ideas inherent in his hyperboloid-based towers into the construction of a double curvature roof. These roofs began to demonstrate the versatility of his applications of non-Euclidean geometry. Where the tower applications were all geometrically similar, although of varying heights and diameters, the applications to roofs allowed for the notion that the diagonal framing typology could support a far wider range of potential geometries.

The lightness and grace of Shukhov's diagrid tower structures may ultimately be their undoing. The exposed nature of the material as well as its choice of lightweight sections and connections has resulted in deterioration. Many are threatened with demolition. The Shukhov Tower Foundation has been established by Shukhov's grandson to spearhead efforts to restore his towers and prevent demolition (www.shukhov.org).

THE IMPACT OF THE MODERN MOVEMENT

The sheer power and popularity of the Modern Movement and the International Style pushed structural experiments and non-rectilinear architectural forms to the side. Although trusses and vaulted and pitched truss-supported roofs continued to form part of building systems that were used throughout the 20th century, the diagonalized grid *per se* simply disappeared. As tall buildings evolved over the century, stability systems preferred the combined use of columns to carry the gravity load and of other methods of non-expressive connection reinforcement to resist lateral forces such as wind loads.

The structural design of steel skyscrapers was reflected in a confidence in high-strength steel in combination with high construction quality that led to the elimination of redundant systems. Bracing systems, particularly those using diagonal members, were eliminated as they interfered with the planning of the interior spaces and increased the floor-to-floor heights. While diagrids provide alternate load paths within their system as a result of the diagonals, regular framing cannot do that. In a description of the 1913 Woolworth Tower in New York City, then the tallest building in the world, the notable architectural historian Carl Condit states:

"The arched frame of the Woolworth Tower extends up to the twenty-eighth floor; above this, to the forty-second floor, bracing is secured through a double system of knee braces in which the knees are located on the top and bottom of the girder and its supporting columns. The Woolworth frame could easily withstand hurricanes of maximum intensity; however, this lavish distribution of steel in deep fillets and braces came to be regarded as an expensive redundancy of metal with an unnecessary sacrifice of the vertical space between the floors. The development of high-strength steel, welded connections and new techniques of riveting and bolting made it possible to eliminate these additional shapes in buildings even higher than the Woolworth."[5]

The Empire State Building, also in New York City and completed in 1931, simplified the type of extensive bracing used in the Woolworth Tower, eliminating much of the redundant steel that was used to reinforce the connections. In this frame, the girders were simply riveted throughout their depth to the supporting column, and the beams riveted throughout their depth to the girders. Triangular braces at the connections were eliminated as these interfered with the clear floor-to-floor height. However, the core structure that housed the elevators and exit stairs was framed in steel, something which has become rare in the meantime and is a typical feature of diagrid buildings. A reduction in floor-to-floor height could allow for increased density on an urban site, meaning more rental income.

Skyscraper construction stayed with this model for decades save for the replacement of rivets by bolts. As building heights were increased and subjected to higher wind loads, new types of bracing systems were needed to reinforce the structure which, in simple terms, had to perform as a very tall cantilever.

Where moment-resisting beam-to-column connections were insufficient, K and X-bracing was added – typically located internally, near the core, in order to make it as unobtrusive as possible; i.e. having no impact on the design of the façade or the flow of traffic in the building.

As requirements for mechanical systems further increased, these were often relegated to designated floors at intervals over the height of the building. Truss structures were used at these floors as a stabilization method. From a design perspective, these truss-band floors could easily be incorporated into the façade planning, while still supporting a standard curtain wall.

GEODESIC DOMES AND SPACEFRAMES

In terms of influence on the concept, structure and detailing of diagrid structures, the invention and development of spaceframe and geodesic structures was important as an overlay to Shukhov's hyperbolic paraboloid towers. Geodesic domes, spaceframes and diagrids all derive their essential stability from triangulation. The detailing of geodesic domes and spaceframes has had an influence on the terminology, engineering and fabrication detailing for diagrids.

As previously mentioned, the technical advances of trusses were important to the development of structural stability in general, as the triangular shape is inherently stable. However, trusses in general were not used to provide support for the overall form or shape of the building; early trusses were developed primarily as planar, two-dimensional spanning members, providing floor or roof support. What is more, trusses developed the concept of a system of connected members that transferred their forces through node points, with the consequence that the design of the nodes primarily served the physical connection of the members and to ensure that the neutral axis of each joining member aligned so as to direct purely tensile or compressive forces through the steel. The nodes were not necessarily governed by considerations of aesthetics or prefabrication and were normally not identified as a clearly fabricated and articulated element.

By contrast, the nodes in geodesic domes and early spaceframes were quite specifically designed to allow for quick, standardized connections and to join a large number of members at a central hub. They were normally prefabricated as physically identifiable objects that formed part of a system. Typically, a structure of this kind has a depth of approximately one module – it gains its rigidity through three-dimensional accumulations of triangles.

Ferrari World Theme Park in Abu Dhabi, UAE, designed by Benoy Architects, uses a fairly traditional form of spaceframe system to create its 205,000m²/2.2 million ft² roof. Member sizes vary according to their loading requirements and the ball-type nodes are designed to accommodate as many as 12 intersecting members.

The Mandarin Hotel in Beijing, China, designed by OMA, shown here during its reconstruction in 2011 (post fire of 2008), illustrates a contemporary variation of a spaceframe application. Where labor is less expensive, it is not uncommon to request welded connections. Consistent with traditional systems-based spaceframe applications, the ball-type joint is capable of receiving a high number of members into its hub. The higher the number of members and therefore more significant the load transfer, the larger the node.

For load transfer purposes, the nodes in trusses, geodesic domes and spaceframes are designed as hinge or pin-type connections. This means that they are not designed to transfer moment forces or to be necessarily stiff. The stiffness of the structure is maintained incrementally through the addition of triangulated sets of members and their nodes. The members tend to be short and therefore relatively easy to lift and self-support during fabrication. Some variations of spaceframes, as in the Mandarin Hotel in Beijing, are tending toward the use of heavier systems with a lesser degree of prefabrication, less consistency in member type and size and using a higher proportion of welding.

The diagonal braces of the John Hancock Center in Chicago, IL, USA, designed by SOM and completed in 1969, clearly express the lateral bracing system as part of the aesthetic of the façade.

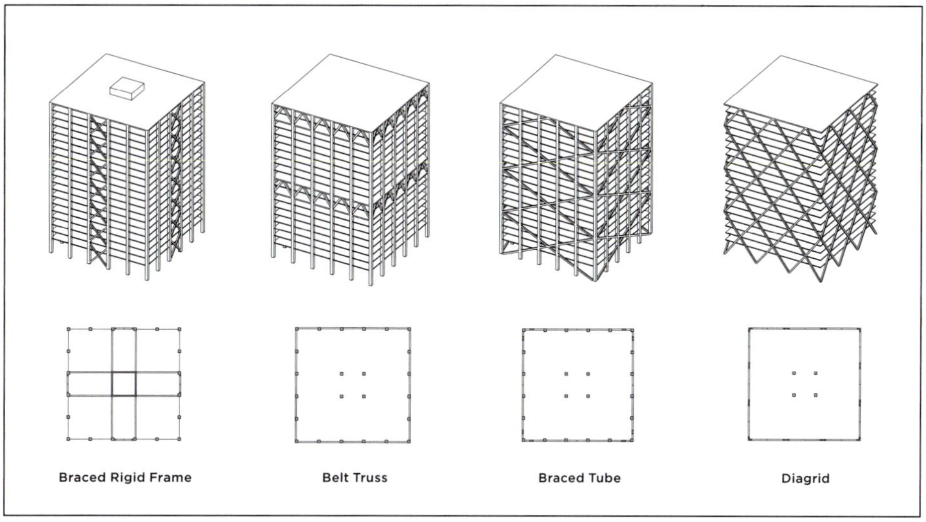

This drawing shows the use of diagonal bracing systems in the high-rise building type. Where the braced rigid frame and the belt truss diagonals are often integrated into a traditional curtain wall framing pattern, both the braced tube system and the diagrid use their scale to transform the design to acknowledge their presence.

THE EMERGENCE OF THE DIAGONALIZED CORE TYPOLOGY

The evolution of the high-rise tower is divided into phases of changing structural framing types. The diagrid is clearly set apart from all other types of structures as the direct result of the elimination of vertical columns as the means to carry gravity loads. The determined expression of its diagonals in the detailing of the façade is another characteristic feature: even in cases where a rectangular curtain wall is applied to the diagrid structure, like with the International Finance Center (IFC) in Guangzhou, the large diagonals of the diagrid remain visible through the highly transparent glass façade.

With the completion in 1969 of the 100-storey John Hancock Center in Chicago, designed by SOM, the braced tubular cantilever was introduced, the first major reappearance of an expressed diagonal brace in a tall building. In order to permit larger expanses of glass and less frequent vertical columns, large diagonal members were overlaid to brace the entire length of the structure.[6] The gravity loads continued to be carried by vertical columns that were framed back to the core, and lateral forces were absorbed by the forcefully expressed diagonal grid. Where previous structures had combined the gravity and lateral force resistance of a column-and-beam system with moment-resisting connections, the John Hancock Center introduced the idea that these two systems could be separated and expressed differently in the architecture.

It was only with the design of the John Hancock Center and the introduction of braced tubes that architects undertook to incorporate the bracing into façade designs, thereby pushing an engineering solution into the realm of architectural design. While previous towers had as a rule maintained uniform plan dimensions throughout the height of the building, the John Hancock Center was tapered. Where earlier designs would set back the tower with steps that were rectilinear in order to decrease the floor area, the diagonal braces introduced angles, which resulted in the modification of the windows at the corners of the building. The John Hancock Center was the first tapered modern tower and served as a precedent for further geometry-related explorations.

The braced rigid frame and belt truss were easily worked into the *status quo* type of curtain wall and hidden behind its rectilinear shape. Architects will express the diagonal bracing nowadays, but when first introduced it was not fashionable to admit to needing to brace the building and its presence was typically repressed in the façade design.

The Bank of China Tower in Hong Kong, China, designed by I.M. Pei and engineered by Leslie E. Robertson Associates, completed in 1990, uses the structure and lines of the diagonal bracing system to support changes in the form of the building.

The Bank of China Tower in Hong Kong, designed by I.M. Pei and engineered by the firm of Leslie E. Robertson Associates in 1990, again made the expressive use of diagonal reinforcing the signature of its design. The 367m/1,205ft-tall building used a composite steel and concrete system to absorb the lateral and gravity loads. The giant diagonal braces are additionally used to modulate the shape of the building, creating triangulated prism-like setbacks. The diagonals are part of the combined wind and gravity system, fabricated as steel box members that were filled with concrete for additional stiffness and damping. The composite material action of the structural elements created a space truss structure that required 50% less steel than typical frame buildings.[7]

The building is a square tube of 170ft/52m on a side, and it is divided by the diagonals into four triangular prisms. The vertical columns are set back from the face of the structure and have no impact on the design of the façade; from the finished exterior it would appear as if there were no vertical columns. The result of the vertical space truss arrangement is that almost all of the gravity loads are transmitted through the diagonals to the four corner columns. A unique innovation in the composite steel and concrete system is the use of the concrete megacolumns at the corners of the building that solve the connection of the giant diagonal braces. Instead of steel-to-steel connections, the ends of the steel braces are embedded in the concrete corner columns (see *Chapter 6: Node and Member Design,* for additional information).

This image of the Bank of China Tower in Hong Kong under construction in 1988 was originally published by *Engineering News Record* in a discussion of the merits of the composite construction system. The hierarchy of the members and the relationship of the large diagonals to the stabilization of the system is evident.

At the time of completion the Bank of China Tower was the tallest building in Asia. There were challenges with respect to the wind environment of Hong Kong, where the wind loads at 143mph/230kmph are twice as high as those of New York City and four times the lateral loading had the building been constructed in the seismic zone of Los Angeles. Pei and Robertson chose this structural expressionist design to reflect this aspect of the design requirements on the form of the building.

Where the John Hancock Center and the Bank of China Tower may vary in the way that the diagonals are used to inform the shapes of the buildings, both designs limit the expression of the diagonals on the façades in order to maintain the use of more standard rectilinear curtain wall systems for the envelope. Different from the John Hancock Center, the Bank of China Tower in its initial square plan has limited the use of vertical columns to the intersection points of the triangulations,[8] so that the curtain walls are uninterrupted by expressions of the vertical columns. The impact of the expression of the diagrid will be examined further in *Chapter 9: Façade Design*.

A FIRST DIAGRID-SUPPORTED OFFICE BUILDING

The first diagrid-supported building stands as an anomaly along the development timeline. The IBM Building, now called the United Steelworkers Building, was completed in 1963 in Pittsburgh, designed by Curtis and Davis and engineered by the firm of Leslie E. Robertson. Robertson was also responsible for the structural design of the diagonally braced Bank of China Tower in Hong Kong, completed in 1990. There is little of the innovative diagrid structural system that is exhibited by the IBM Building that would be seen to directly influence Robertson's later work. However, the prefabricated nature of the elements and the strength of the perimeter tube directly influenced the design of the exterior structural system for the World Trade Center Towers in New York City.

The diagrid on the IBM Building serves the simultaneous functions of structural support system and cladding. For its construction time in the 1960s, this must have seemed very odd as the classic aluminum curtain wall was being used extensively in high-rise construction, based on the separation of these two functions established since the late 1800s. The load of the exterior comes to the ground at only eight points - two on each side of the building. This creates significant cantilevers from this pair of central supports to the corners of the building. The structural steel diagrid is clad to prevent corrosion.

Although the widely spaced support points at grade create a very large cantilever at each of the four corners of the building, the modularity of the diagrid is very fine. The diamond grid effectively uses one diamond-shaped module for each floor-to-floor height, placing the widest point of the diamond that frames the vision glazing at eye level. By contrast, unlike in future diagrid buildings, the entire structural system is not triangulated. Horizontal tie beams only occur at the floor levels where the steel structure attaches to the floor, so that the vision from inside remains unobstructed.

This image of the IBM Building under construction clearly shows the light nature of the diagrid as an external support system for the building. The firm of Leslie E. Robertson notes that the term "diagrid" was not in use at the time of construction, but that this is considered to be the first modern use of the system. The diagrid frames were color-coded according to their strength requirements for assembly.

The IBM Building in Pittsburgh, PA, USA (now United Steelworkers Building), designed by Curtis and Davis and engineered by Leslie E. Robertson, completed in 1963, is the first example of a steel diagrid structural system that has been used to support the floors of a tower building. Departing from the classic aluminum curtain wall system, the steel diagrid exoskeleton assists in stability and is clearly expressed in the façade.

The integration of the glazing into the diagrid structure creates a more expensive cladding solution, as the hexagonal-shaped windows and infill panels are non-standard geometries that required specialty fabrication and erection. The construction of the cladding system was an effective way of providing corrosion protection for the steel structure – the building is in extremely good shape and very well maintained as of 2013.

Although the prefabricated system created for the IBM Building was never precisely replicated in later diagrid buildings, this building did set out the primary strategy that is currently used: a perimeter diagrid tube frame that combines the lateral and gravity load-bearing systems. The vertical columns on the exterior were eliminated as well as the interior columns, leaving a very open floor area. The exterior diagrid was connected to the core via a steel-framed floor system. The core was framed in steel but did not have the same level of reinforcement as previous tall buildings, as the lateral loading was resisted by the perimeter diagrid. This diagrid system differs from a traditional perimeter tube system in the elimination of the vertical column system for gravity loads. The diagrid tube system simultaneously resists gravity and lateral loads, so that the assistance of the central core is in principle no longer indispensable. The specific requirements of the core in load resistance will be discussed in detail in *Chapter 7: Core Design*.

The cladding of the steel diagrid creates a very fine grain for the façade aesthetic. The coloration of the vision glass is quite close to that of the solid spandrel panel infill elements.

The exterior diagrid frame of the IBM Building is clad for corrosion protection. A concrete waffle slab is used to assist with supporting the large cantilever to the corners of the building.

The diagrid for the IBM Building was prefabricated into modules for ease of site assembly, with primary structural connections occurring at the mid-points between the floor levels.

THE FORMATION OF THE CONTEMPORARY DIAGRID STRUCTURE

In retrospect, several key factors contributed to the emergence of the contemporary diagrid structure, which at the same time mark the key phases of transition from Shukhov's hyperbolic paraboloid towers:

- → Where Shukhov allowed the members of his diagrid towers to simply overlap and often be continuous through the "theoretical nodes", geodesics and spaceframes introduced the technique of the nodal connection, thereby making the members discontinuous at these critical points of connection.
- → Diagrid structures make use of prefabrication and erection techniques that became highly developed in the fabrication of spaceframe structures (although there are obvious differences due to the increase in scale of the diagrid system and its members).

The combined scale and load increase in contemporary diagrid structures shifted the structural function of the node as a true pin or hinge connection to one that required moment resistance. We will further examine moment resistance in *Chapter 6: Node and Member Design*. Moment resistance, as we will see in *Chapter 8: Constructability*, is also necessary to reduce the need for temporary support systems during erection, so that the members can cantilever with little deflection from the node in anticipation of further connections to complete their triangulation.

The use of a moment-resisting or stiffened node system to connect the straight members of the diagrid transformed the original perimeter support method established by Shukhov into a combined lateral and gravity load system that was capable of supporting buildings with significant loads. The increase in the scale of the members allowed for the introduction of a floor system (that was not part of the Shukhov type) that could span from the perimeter diagrid system to the core.

The placement of the primary diagrid tubular structure at the perimeter of the building, as first introduced in the IBM Building, combined with expressed bracing as first introduced in the John Hancock Center and since increased in scale, set the stage for systematic refinements leading to the emergence of a defined diagrid structural system.

A TIME OF STRUCTURAL CHOICE

The contemporary creation of convoluted geometries and curved forms strongly suggests that a triangulated approach in steel may provide highly appropriate structural solutions. For larger buildings that support multiple floors this may take the form of a diagrid structure. However, it is by no means suggested that this is the only or even necessarily the best solution for the construction of all contemporary large buildings. For roofs or shell-type structures, a lattice grid may be more appropriate (see *Understanding Steel Design, Chapter 12*). Developments in 3D structural modeling have made these options very viable. For tall buildings that support significant gravity and lateral loads there are a number of strategies to choose from.

A **diagonalized core structure** continues to be a popular choice for tall buildings. A characteristic of this type is the larger size for the diagonals, creating modules that span up to 12 floors as a function of the overall dimensions of the floor plates and the height and shape of the tower. The diagonalized core system uses the diagonal members to supplement the vertical load paths provided by columns. The cross-sectional area of the diagonals is significant and they can have a large impact on the design of the façade as well as on the planning of the interior spaces.

The diagonalized core system is sometimes used in conjunction with a newer composite framing type that employs **megacolumns**, especially in Supertall towers where the cumulative floor loads are very high. A Supertall tower is defined by the Council on Tall Buildings and Urban Habitat (CTBUH) as having a height of at least 300m/984ft. Megacolumns can reduce the frequency of the columns, sometimes down to eight, but are also dependent on the core for stability. As large diagonals can obliterate the view, a combination of a diagonalized core system with megacolumns may also keep the relative size of the diagonals down, as in the KK100 Tower in Shenzhen, designed by TFP Architects with Arup, a Supertall measuring 442m/1,499ft.

The structural system for the Chow Tai Fook (CTF) Tower in Guangzhou, China, designed by Kohn Pederson Fox and Associates with Arup, uses a megacolumn strategy. This photo of 2012 shows the relative size of the columns and the reinforced concrete core on the lowest levels of the future 530m/1,739ft Supertall tower.

Megacolumns and also megaframes are even newer to the structural scene than diagrids. For Supertall towers such as the Shanghai Tower, designed by Gensler Architects with Thornton Tomasetti Inc. Engineers, a **core-and-outrigger megaframe** is used to meet Chinese building regulations. This tower is projected to reach 632m/2,073ft, i.e. close to 200m/656ft more than the Guangzhou International Finance Center (IFC), which is the tallest diagrid tower constructed to the time of writing.

This site model for the KK100 Tower in Shenzhen, China, designed by TFP Architects with Arup, shows the use of a diagonalized core on a Supertall building. The photo was taken inside KK100 in the presentation center. The actual size of a diagonal can be seen in the background.

Guangzhou Fortune Center is a diagonalized core building in Guangzhou, China. This image of the future 309m/1,015ft Supertall, under construction in 2012, illustrates the use of custom box sections that will be filled with concrete to create the steel structure, which is a very common construction method in China. The splices on the columns and diagonals are welded. The structure will be concealed. The core is reinforced concrete.

Whether or not a diagrid framing system is a structurally viable solution for this category of Supertall towers has yet to be decided. The diagrid versions of the designs for the Zhongguo Zun Tower in Beijing by TFP Architects (528m/1,732ft, 108 floors) *(see Project Profile, p. 170)* and the Lotte Super Tower in Seoul by SOM (555m/1,819ft, 123 floors) *(see Project Profile, p.172)* have both been abandoned in favor of other structural solutions.

Some of the planning issues when selecting the appropriate structural system are the impact that the structure has on the interior spaces, and how to use the structural system for architectural advantage or purpose. **Outrigger systems** will often make use of the outrigger floors as places of refuge in response to evacuation and fire safety needs. These are normally expressed on the façade by a change in the curtain wall design and materiality. Megacolumns have a large impact on the usable floor area of the building, particularly on the lower floors.

In the end, the structural system choice requires the collaborative evaluation of all conceivable design options by the project team.

The base of the Shanghai Tower, Shanghai, China, designed by Gensler Architects with Thornton Tomasetti Inc. Engineers, under construction in 2012, reveals the complex curved cladding that is part of a double façade system hung from the outrigger floors. The megacolumns of the core-and-outrigger megaframe are visible at the base of the building.

A view inside one of the outrigger floors of the Shanghai Tower. These very large-scale trusses are tied back to the concrete core in order to be able to support the floor and tension system that is cantilevered from the megacolumns. This complex solution facilitates the irregularly curved form of the tower.

The size of a megacolumn for a 632m/2,073ft Supertall tower is extremely large. Note the scale of the construction worker standing beside one of the base columns for the 128-floor Shanghai Tower.

3

PRINCIPLES OF THE CONTEMPORARY DIAGRID STRUCTURE

CONCEPT AND DEFINITION

EXPLORING THE POSSIBILITIES OF DIAGRID SYSTEMS

MATERIAL CHOICES

STRUCTURAL BENEFITS

THE FIRST CONTEMPORARY DIAGRID BUILDINGS

—PROJECT PROFILE: LONDON CITY HALL | FOSTER + PARTNERS, ARUP

—PROJECT PROFILE: SWISS RE TOWER | FOSTER + PARTNERS, ARUP

—PROJECT PROFILE: HEARST MAGAZINE TOWER | FOSTER + PARTNERS, WSP CANTOR SEINUK

The Hearst Magazine Tower, completed in 2006, in New York City.

CONCEPT AND DEFINITION

The term "diagrid" is somewhat misleading. Diagrid is commonly used to describe a diagonal structural grid. The system is comprised of diagonal members, normally fabricated from structural steel, that are joined at nodal points. The diagonal grid, although often presented as the dominant visual feature in the design of diagrid buildings, is by itself unstable. The diamond-shaped system requires triangulation in order to create sufficiency in the structure. Diagrids or diagonal grids are a structural design strategy for constructing buildings that combine the resistance to gravity and lateral loads into a triangulated system of members that eliminates the need for vertical columns. This system is usually placed on the perimeter of the building. Triangulation is normally achieved where the floor edge beams tie into the grid.

The primary idea behind the development of the diagrid system was the recognition of the savings possible in the removal of (most of) the vertical columns. Vertical columns, engineered to carry gravity loads, are incapable of providing lateral stability. The diagonal grid, if properly spaced, is capable of assuming all of the gravity loads as well as providing lateral stability due to its triangular configuration. A pure diagrid structure does not require the traditional reinforced concrete or steel core to provide lateral stability. This will be further addressed in *Chapter 7: Core Design*.

A diagrid tower is modeled as a vertical cantilever. The size of the modules of the diagonal grid is normally determined by evenly dividing the height of the tower. Numerous studies have been conducted toward the optimization of the module size as a function of the building height and angles of the inclined members.[1] The diamond-shaped modules typically span six to eight floors tip to tip, although shorter modules are used for buildings with irregular geometries or tighter curves. Normally, the height of the base module of the diamond grid will extend over several storeys. In this way the beams that define the edge of the floors can also frame into the diagonal members, providing the required stiffness to the unsupported length of the diagonal members. Floor framing will usually connect these edge beams back to the core.

As a significant portion of the expense of the structure lies in the fabrication of the nodes, efforts are toward minimizing their variation and frequency and simplifying the connection between the node and the diagonal to speed up erection.

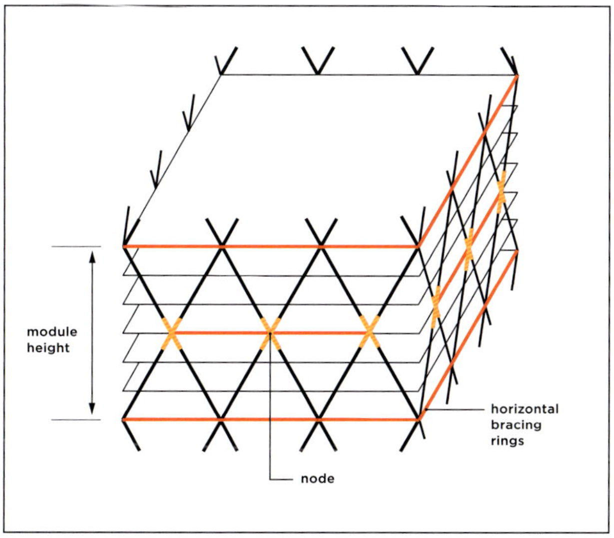

Attributes and terms of the diagrid. The "module" refers to the number of floors that the diamond shape of the grid spans from tip to tip. The "node" is the point of intersection of the diagonal members. A "horizontal bracing ring" is created by the connection of the diagrid nodes to the floor edge beam. The steepness of the angle of the diagrid is measured as the angle formed between the diagonal and the floor. This diagram illustrates the application of the system to a rectangular plan. This can result in large cantilevers of the corner floor areas, as the corners are without vertical columns to assume their support. (Diagram after K.S. Moon, see note 1.)

The major structural intersections occur at the nodes. The diagonal members are typically continuous from node to node-conversely inferring that they are erected as discrete members as there is a physical connection between the member and the node. This will vary from project to project as a function of the module height.

The diamond-shaped modules must be braced at the very least at their widest point, using a node-to-node connection to complete their basic structural triangulation. Depending on the overall geometry of the building, the horizontal bracing or rings can be required to act in tension (where the gravity loads would cause the diagrid to push outwards) or in compression (where the slope of the diagrid would push inwards on the building). The horizontal brace is often formed by the edge beam of the floor structure, which frames into the node to complete the triangle. Alternatively, an additional expressed structural steel member can be fixed between the nodes.This will be discussed further in *Chapter 5: Modules and Modularity* and *Chapter 6: Node and Member Design*.

EXPLORING THE POSSIBILITIES OF DIAGRID SYSTEMS

Where early applications of expressed diagonal bracing tended not to significantly modify the basic rectilinear shape of the tower (John Hancock Center in Chicago, for example), current applications of the diagrid are exploiting the ability of the triangulated mesh to more easily distort and create curved or more random geometric forms. The term "mesh" makes direct reference to the mapping techniques of 3D modeling software and the means to make fairly direct translations from architecturally driven design investigations, through BIM to fabrication detailing software. The notion of a mesh that is fitted to complex curves is the basis of the design of contemporary lattice grids.

Most diagrid towers work toward the elimination of columns between the exterior structure and the core. The more striking diagrid examples also tend to strive for effective daylighting and use a small floor plate. This supports a sustainability-motivated trend toward increased daylight effectiveness and LEED™ credits.

What is incredibly intriguing with diagrids, in contrast to previous Modernist structural strategies for tall buildings, is that a "basic" typology does not exist. Contemporary diagrid buildings tend to be unique, reinforcing the idea that architectural ambitions are pushing the engineering technology in these structures. Whether in terms of height, shape, profile, node design or the length of the diagrid member, each and every diagrid structure is very different, almost in defiance of current research looking for optimization. This can be attributed to advances in computing and modeling that have run parallel to diagrid development, if not slightly ahead, easily supporting curvilinear or irregular geometries throughout the design process from structural design to detailing, the creation of shop drawings and fabrication.

MATERIAL CHOICES

Although a diagrid could be created using a number of different materials, the most often used structural material is steel. The steel diagrid perimeter framing system can be used in conjunction with a steel-framed or concrete core. The choice of core material will be discussed further in *Chapter 7: Core Design*. A combination of a steel diagrid with a steel core would be considered an all-steel structure according to the Council on Tall Buildings and Urban Habitat (CTBUH), while a steel diagrid in combination with a concrete core would be classed as a composite structure. The use of a steel diagrid system comes at a point in history when the use of steel for the construction of tall buildings is at an historic low. According to CTBUH data only 2% of all tall buildings (200m /656ft or higher) constructed in 2012 were created as all-steel buildings. This is down from 3% in 2011, 22% in 1992 and 63% in 1972.

Where the choice of the core material may vary by locality, the selection of steel for the diagrid system itself is universal when it comes to the fabrication of a system of elements that lends itself to mass fabrication, the use of Building Intelligence Modeling (BIM) and faster on-site assembly. Concrete does not lend itself to this sort of mass fabrication. Another reason is that many of the elements must resist high tensile forces.

Studies by engineer Barry Charnish for the design of the Bow Encana Tower in Calgary (see Process Profile in *Understanding Steel Design*, pp. 136–143), completed in 2012, concluded that a 20% savings in the weight of the structural steel was possible using a diagrid/hybrid structure versus a braced tube structure. Reports also claim that 20% less steel was used in constructing the Hearst Magazine Tower in New York City using a diagrid structure over a conventionally framed steel structure.

STRUCTURAL BENEFITS

There are a number of structural advantages that can be attributed to the use of a diagrid system in comparison to a typical orthogonally (moment-) framed system. Where the early "diagonalized core" or "braced tube" system used diagonal bracing members over a framed exterior support system as a supplementary stability system, the current diagrid system often uses an exclusive exterior frame comprised entirely of diagonal members as the primary means of combined gravity and lateral support. The perimeter diagrid often cooperates with the central core that contributes to carrying the gravity loads and provides additional stiffness. This type of structure carries lateral wind loads more efficiently than vertical columns, creating stiffness that is complemented by the axial action of the diagonal members:

"Compared with conventional framed tubular structures without diagonals, diagrid structures are much more effective in minimizing shear deformation because they carry shear by axial action of the diagonal members, while conventional framed tubular structures carry shear by the bending of the vertical columns."[2]

The diagrid obviates the need for large corner columns as is the case in the Hearst Magazine Tower in New York City and in the IBM Building in Pittsburgh, where the building is supported only on eight points, leaving the corners completely cantilevered. The diagrid also provides a better distribution of load in the case of a compromised building as the system creates redundancy in handling load paths as a direct result of the diagrid action.

Where recent Supertall buildings such as the Burj Khalifa in Dubai, the proposed Kingdom Tower in Jeddah and the China Broad Group "Sky City" Tower in Changsha[3] increase the size of the building base to resist moment, most diagrid towers have not deployed this tactic to building reasonably tall. Some, such as Swiss Re and the Guangzhou International Finance Center have even narrowed the base and rely on the diagrid structure for added stability. The tallest diagrid tower constructed to date is the Guangzhou IFC, designed by Wilkinson Eyre Architects with engineering by Arup. At 439m/1,439ft it qualifies as a Supertall building, but is well short of the record set by the Burj Khalifa (828m/2,717ft), which will soon be surpassed by the Kingdom and China Broad Group towers. However, the relatively straight tubular design of diagrid towers permits significantly closer spacing between buildings, which is an important issue if looking for urban density.

THE FIRST CONTEMPORARY DIAGRID BUILDINGS

The contemporary diagrid building made its appearance in the early 2000s as three projects were almost simultaneously underway in the offices of Foster + Partners. London City Hall was completed in 2002. Swiss Re (also known as 30 St. Mary Axe or "The Gherkin"), also in London, commenced design in 1997, started construction in 2000 and was completed in 2004. The Hearst Magazine Tower in New York City broke ground in 2003 and was completed in 2006. The engineering expertise of Arup was integral to the design and detailing of the London City Hall and Swiss Re projects. The engineering firm of WSP Cantor Seinuk collaborated on the Hearst Magazine Tower.

PROJECT PROFILE: LONDON CITY HALL, LONDON, ENGLAND

London City Hall represents the collaborative efforts of Foster + Partners and Arup with respect to its sustainability agenda as well as its novel structural form. The geometry of the 45m/148ft-tall building very directly responds to climate and solar issues.

ARCHITECT
FOSTER + PARTNERS

STRUCTURAL ENGINEER
ARUP

M+E ENGINEER
ARUP

CONSTRUCTION MANAGERS
MACE

The curved shape of the north face of the building uses a diagrid structure, leaving the south side more conventionally framed using inclined columns to accommodate the varying angles of the oval stepped floor plates. The diagrid is comprised of circular hollow structural sections (HSS) made of Architecturally Exposed Structural Steel (AESS). As this structure would support and house the high-profile spiral ramp in close viewing range, it was essential that the quality of the steel would be exceptional. The structural steel used for the office space was simpler, with the inclined columns made of AESS and expressed in the finished space providing a unifying element.

London City Hall in London, England, designed by Foster + Partners with Arup and completed in 2002. The structure leans away from the water to form a stepped façade on the south side to create sustainable self-shading.

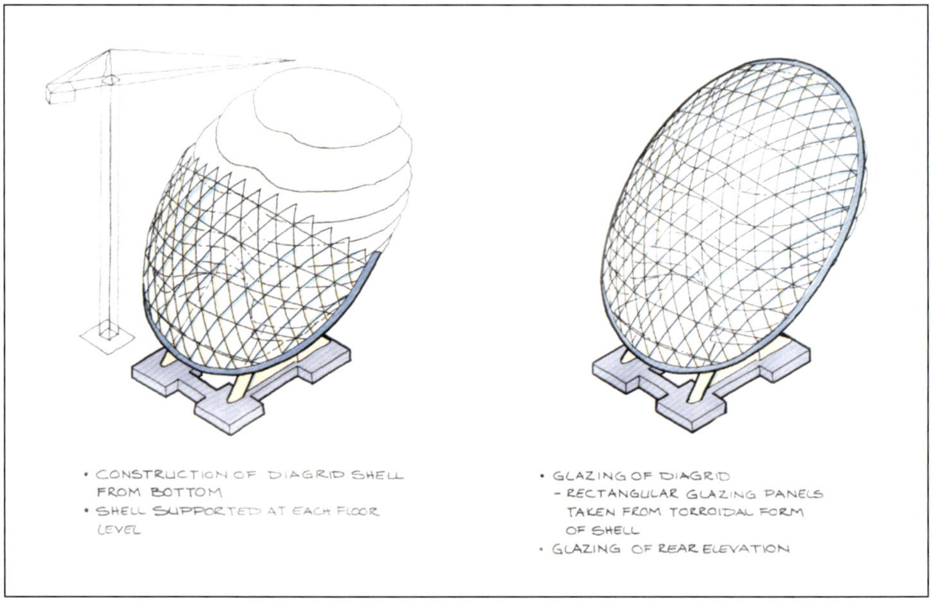

These drawings produced by Arup were part of a study to set out the parameters for the diagrid structure. The geometry of the egg shape is fitted to the floor plates. The void through which the spiral ramp will run can be seen behind the façade.

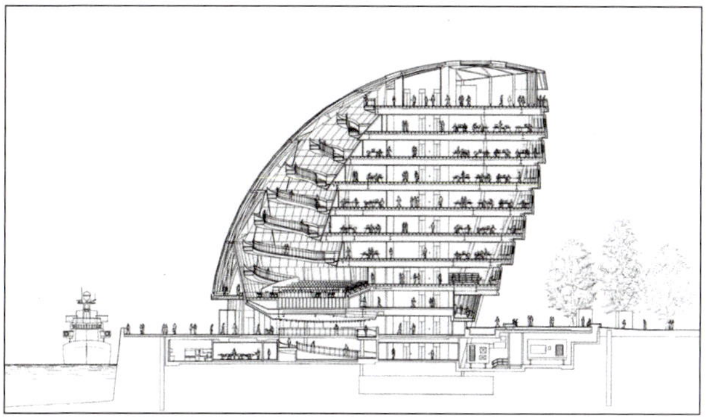

This cross-sectional view illustrates the separation of the building into the north-facing zone that contains the spiral ramp and the southern side that contains the office spaces.

The diagrid shell, although rigid itself, needed to be tied back to the floors of the building (as noted in the Arup diagram) as it formed only part of the structure. In essence, it was primarily designed to support the curved glazing as well as to lend support to the large internal spiral ramp as it passes by the façade.

A structural challenge for any building is the creation of a large, eccentric void. The London City Hall design has segregated the ramp, relying more completely on a system of tensile supports instead of hard connections for its support. The stiffness of the diagrid shell is required to allow the spiral ramp to be suspended from the structure using a system of tension rods.

The top image of 2001 of the City Hall during construction shows quite clearly the different approaches taken to structuring the diagrid and the occupied office floors of the building.

A view from the public access ramp to the private ramp that spirals up above the Council Chamber. The central spiraling ramp is stabilized via tension rods that are tied back to the diagrid above.

The module of the diagrid does not correspond with the floor height to span an even number of floors, as the connection points of the diagrid are not cleanly aligned to office floor levels. Clearly visible from the interior, the alignment of the grid is set slightly above the floor level. The base of the diagrid is expressed and effectively serves as a railing or bench at the exterior side of the building. The module height is approximately four floors, with each diagonal spanning two floors. As the entire structure, is expressed on the interior, the visible horizontal braces that complete the triangulation of the diamond grid have been created with similar round hollow structural sections to achieve a unified appearance. We will see in future examples that the horizontals are more normally buried in the floor edge structure, but as these were raised above the floor they needed to be treated in the same way as the diagonals to make a very coherent-looking exposed system. The construction image shown above shows the attachment of connectors for the glazing at third points along the diagrid members. The detail photo shows how the connectors attach to the steel lattice grid that forms the direct support for the exterior, triangular glazing.

The fire engineering approach[4] comprised the fire protection of the exposed steel columns with a thin-film intumescent coating, providing a high-quality finish. The ability to employ thin-film intumescent coating supported the type of detailing selected for the diagrid. The decision for all-welded connections in this AESS diagrid application will have increased the cost of fabrication and erection, as the amount of shop welding prior to shipping will have been limited - shop welding being preferable as it is easier for the fabricators to manipulate the steel for better access to the welding. As a result this diagrid structure does not create distinct nodal elements to which the diagonals would be connected. Such nodal elements have become the normal procedure for much larger projects as they are very advantageous in fabrication and erection of the structure.

The main public gallery with the diagrid structure to the left and the glazed overlook to the Council Chamber on the right.

Base of the diagrid structure in the public gallery. As the horizontal elements form an integral part of the expression of the exposed steel, they have been fabricated from the same circular hollow structural sections as the diagonal elements. The all-welded nature of the structure is clearly evident. The welds have been cleanly done so that no grinding or remediation has been deemed necessary. This is a great cost saving to AESS projects.

Given the curved shape of the building at a relatively small scale, it was necessary to curve the diagrid members that span over two floors from bracing point to bracing point. On larger-scale buildings it is possible to achieve the impression of curvature by using straight members. The close viewing contact with the steel is another factor that would not tolerate such an approximation. By contrast, the triangular glazing panels are supported by straight steel segments, as there are three glazing segments to each diagrid member so that the effect of the curvature is not a factor here. Maintaining straight edges within such a curtain wall structure is important, as the sealed glazing panels are planar in nature and curved glazing, particularly for sealed units, is prohibitively expensive.

Two features of the exterior façade finishing highlight the diagrid on the front of the building. The use of straight glazing support sections lend the triangulated façade a slightly faceted appearance. The office floors are set off by smaller rectangular glazing units. These make the incorporation of operable window units and internal shades more cost-effective. The position of the diagrid-enclosed spiral ramp on the north side of the building did not require a shading system.

Although London City Hall has become an integral and vital architectural player in the "New London", its form was not readily accepted on completion in 2002. Don Barker, a writer for the US-based online journal *ArchitectureWeek*, described it as looking more like a "moon base landing unit". In retrospect, the project is important for its early explorations in the use of the AESS-constructed diagrid in conjunction with its irregular curvilinear form. Very different from Swiss Re and the Hearst Magazine Tower, which were both in process during the construction phase of London City Hall, it did provide early experiences with some of the unique design and construction issues associated with the diagrid system.

At the point where the diagrid façade meets the steel lattice-type support for the glazing (top left), the latter aligns with one triangular module of the diagrid. In the diagrid structure, connecting plate tabs are inserted between the glazing sections. Each glazing panel consists of nine triangular panes. The exterior expression of the prefabricated glazing panels is evidenced by a slight thickening of the black mullions where they align and connect to the diagrid. This suppresses rather than accentuates the modularity of the panels.

A plate tab welded to the circular hollow steel section (top right) facilitates the attachment of the curtain wall. Departing from the conventional aluminum curtain wall type, the structural system to which the glazing is attached has been fabricated from plate steel effectively acting as a lattice grid.

The diagrid frame has been intricately set into the overall façade of the building (left). Its form clearly expresses the modularity of the diagrid derived from the diagonals aligning with two floors. The length and geometry of the elements vary over the height of the building: the diagrid that spans over the top two floors was somewhat stretched to accommodate the changing dimensions.

PROJECT PROFILE: SWISS RE TOWER, LONDON, ENGLAND

ARCHITECT
FOSTER + PARTNERS

STRUCTURAL ENGINEER
ARUP

WIND ENGINEERS
RWDI

FIRE ENGINEERING
ARUP FIRE

MAIN CONTRACTOR
SKANSKA

STEEL CONSTRUCTOR
VICTOR BUYCK STEEL

CONSTRUCTION
HOLLANDIA

FAÇADE CONSULTANT
EMMER PFENNINGER PARTNER AG

FAÇADE SUPPLIER
SCHMIDLIN (UK) LTD.

The Swiss Re Tower, also known as 30 St. Mary Axe, is the first full diagrid office tower to be completed. It has defined the high-rise diagrid building typology, including the key elements of module and node. The thorough parametric engineering and wind studies carried out for this project by Foster + Partners, Arup and RWDI established the basis of subsequent diagrid tower design. It is a prime example of the requirements for a collaborative team approach between the architectural practice, engineering firms and steel contractors as the basis for ensuring a successful outcome when undertaking such a unique initiative. Although Foster has been quoted as referencing the work of Shukhov as an inspirational precedent for Swiss Re, the physical realities and requirements of the projects are extremely far apart. From Shukhov's hollow, open framework tower to Foster's idea for a highly energy-efficient double façade building that uses a spiral-based strategy to encourage natural ventilation, there was a lot of technical ground to be covered.

Some of the technical innovations of the London City Hall project were helpful to inform the design of Swiss Re, but even these projects were drastically different in terms of their height and programmatic requirements.

One of the key challenges to providing engineering input for this building was, according to Arup, the requirement to design with varying geometry. Arup developed the diagrid perimeter structural form with Foster + Partners very early in the design process, in order to handle the combined geometry and program requirements. The specific nature of this diagrid was very much tied to the circular form of the structure and the desire for the spiraling double façade ventilation system. Dominic Munro, a design associate with Arup who was deeply involved with the project, wrote in an article for *Stålbyggnad*: "The project shows the ability of structural steel to enable radical architectural ideas to be realized."[5]

The site context and the desire to open up a public plaza contributed to the development of the tower's circular plan. The diameter was varied so that the footprint of the building on the site could be smaller, with the tower widening to 56.15m/184.22ft at the 17th floor and narrowing again to the top at 179.8m/590ft. As a result, each of the 40 floors above ground is different. The floor span, orientation and the intersection angle of the floor with the perimeter façade all vary throughout the height of the building. According to Dominic Munro,[6] "the perimeter steel structural solution was developed specifically for this building in order to address the issues generated by the unusual geometry in a manner that was fully integrated with the architectural concept..."

Swiss Re Tower (also called 30 St. Mary Axe) in London, England, designed by Foster + Partners and engineered by Arup, completed in 2004, was the most unique tower on the Central London skyline for several years. Its height was recently surpassed by The Shard (from where this photo was taken). The Leadenhall Building by Rogers Stirk Harbour + Partners (see *Project Profile*, page 148) is encroaching on Swiss Re's spot on the London skyline.

The floor plan diagram shows the layout of the internal space conceived as six fingers arranged around a central circular core hub. The diagrid structure brought about the substantial benefit that the core was not needed to resist wind forces and could be designed as a steel structure based on an open plan to provide flexible space planning. Transferring the responsibility for wind resistance and lateral stability from the core to the perimeter diagrid also meant that foundation loads were reduced compared with a building stabilized by the core.

In order to keep the floor areas column-free, the columns were placed in the perimeter double façade zone that is situated between the exterior curtain wall glazing system and the interior glazing system. Then the perimeter "columns" were greatly inclined as they became the diagonal members of the diagrid system. As a consequence they follow the tapering shape of the building and, as the diagrid is symmetrical in its conception, this resulted in the creation of spirals that run up the building in opposing directions.

From the six-finger layout of the interior office areas emerged six triangular voids that are used as part of the double façade ventilation system and also serve as light wells to the interior spaces. Essentially, the floor plan was twisted or rotated at every floor level to follow the natural triangulation of the tapering diagrid. The office spaces spiral up the building, following the helical lines of the diagrid. This in turn avoided the need for large cantilevers in the floor slabs and balanced the diagrid structure.

The relationship between the diagrid and the floor slabs can be seen in this construction image. The nodes are located at alternate floor levels, meaning that the tubular steel diagonals span two floors. Where they span across a ventilation shaft, they are not braced. The floors tie into the diagonals at the edges of the occupied interior space. Where the diagrid needs to resist circumferential stresses, the mid-height "hoops" use more slender structural sections as they primarily serve as tensile elements.

If technically the "columns" of the perimeter are created by the diagonal members of the diagrid, connected at alternating floors by the "nodes", another system was required to provide horizontal restraint to the system, as the gravity loads naturally act to push the diagonals outward. Arup added horizontal "hoops", which connect to the column intersection points (nodes) and resist the forces arising from the curved shape. The three-dimensional geometry arising at the nodes, due to the inclined nature of the columns, results in very high outward thrust around the mid-height of the building. Essentially, the hoops act as tension rings for the middle and lower sections of the tower. At the top of the tower, where the loads are lighter and the diagonals more steeply inclined, the rings act in compression. The hoops turn the diagrid into a very stiff triangulated shell, which provides excellent stability for the tower. The hoops also work to resist any asymmetrical or horizontal loading conditions. The wind engineering of this project will be discussed in *Chapter 4: Technical Requirements.*

The curved building structure is created by using straight segments. Unlike the diagrid structure for London City Hall, the scale of Swiss Re is large enough to allow for a faceted structure that results in a curved appearance without the added complication and expense of bending the steel. The further subdivision of the triangulated diagrid pattern in the façade system assists in creating the illusion of the curved surface.

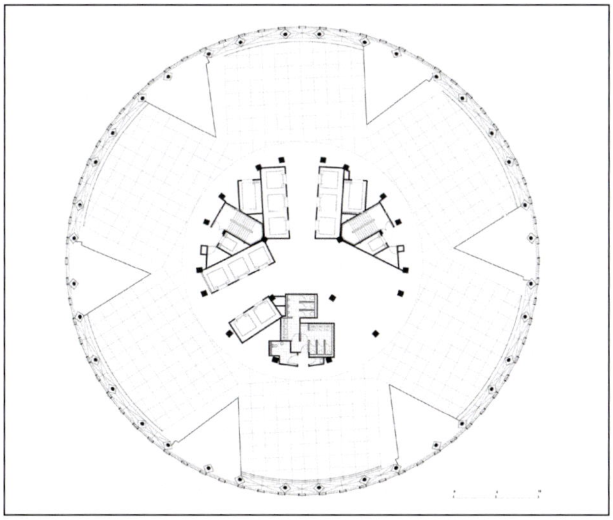

A mid-level floor plan of Swiss Re illustrates how the six-finger plan relates to the triangular-shaped voids that serve to facilitate ventilation in the building. The core is simply structured with steel columns. The diagrid elements are positioned within the double façade envelope at the point where the office space touches the exterior.

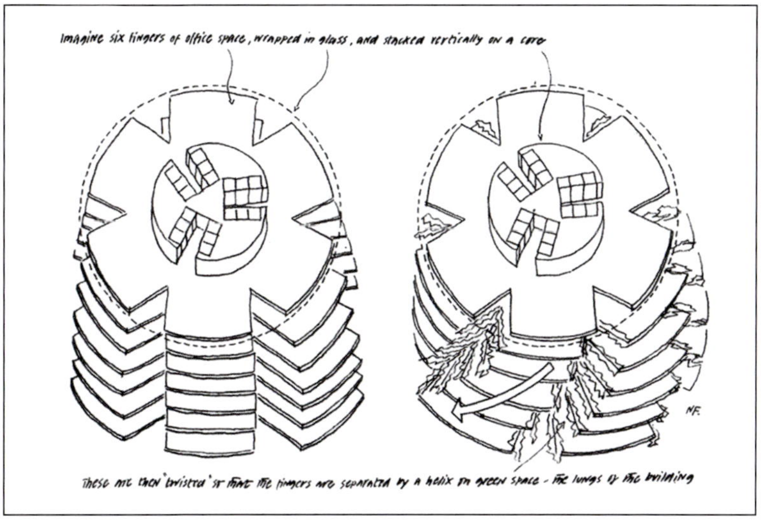

This sketch from the architect illustrates the way the floor plates rotate to create the spiraling triangular voids. The architectural intention was to keep the view through the spiraling glazed ventilation shafts clear so that the occupants would have a visual connection across the space.

The integration of the structural design with the architectural and environmental design intentions necessitated a highly collaborative approach. Digital tools were critical to the success of the project. Arup's approach to the structural analysis, design, coordination and the communication of construction information included the development of a range of fitted parametric design tools customized to the specific geometric form of the building. As this sort of software and working method was relatively new at the time, it was important to develop internal links between the analysis, design and 3D steelwork modeling software. Arup used a full Xsteel model for this purpose. One of the project initiatives was to exchange electronic information freely within the design team and with the contractors. The 3D model proved to be indispensable for effective project communication: while the structural engineer created the initial coordination model with center-lines and sizing, the contractor and subcontractors used this model for detailing and interfaces with cladding and MEP services.

The three-dimensional nature of the design considerations that followed from the diagrid façade required novel solutions. Tubular sections were chosen to master the twisting geometry between diagrid nodes. Round plates that were welded to the ends of the diagrid members allowed for an easy bolted connection with the nodes. The circular shape of the diagrid members made the sections much less bulky in comparison to a Universal section or a wide flange profile. This was important to the design as the fire engineering for the project was challenging.[7]

Sub-assembly of diagrid members to the nodes can take place on the site to minimize site work in the proper sense to some degree. The node and two members are lifted as a unit as shown to keep the structure stable during erection.

Unlike a standard rectilinear high-rise building that might treat the structural frame in a less celebrated manner, this premier diagrid tower intended to express the structure. The ultimate decision of the fire risk analysis required a 90-minute protection on the structural steel. It was decided to clad the steel as part of the fire protection process. The round steel sections allowed for the diamond-shaped "boxes" to be rotated in such a way as to make minimal impact on the façade. In tune with the minimal engagement of the diagrid with the curtain wall, only the color of the mullion (silver), not its thickness, is modified on the curtain wall to acknowledge the placement of the diagrid behind the curtain wall mullion system.

Swiss Re was important in the establishment of the nowadays rather standardized use of nodes in combination with straight diagrid members. This arrangement greatly simplified the fabrication and erection of diagrid structures. Node design

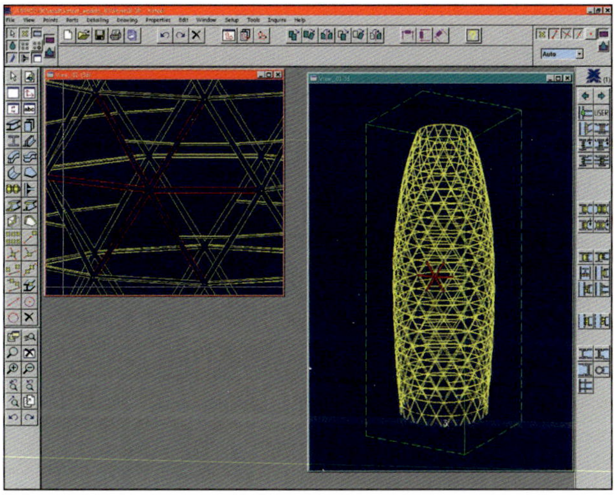

A screenshot of one of the digital working models from Arup.

A detailed screen view of the development of the digital structural model showing the refinement of the node and the connection to the incoming radial floor members.

was a challenge as the precise geometry of the nodes changed for each floor as a function of the curve of the building. Generally, the nodes appear the same, besides adjustments in the angle. The accuracy of fabrication of the prepared bearing surfaces of the nodes and columns was critical, requiring milling to a tolerance of 0.1mm/0.004in. This ensured a very good level of fit with minimal site adjustment needed. Tight tolerances were also required as not all of the steel was fabricated in the same shop and would not come together until erected on site. Typically a bolted connection was chosen to facilitate quick assembly of the incoming diagonals, while the nodes' internal elements were shop-welded.

The team also needed to devise a strategy for accommodating the natural horizontal spread of the diagrid during the erection process, while simultaneously making a connection to the floor. The steel contractors VB-H developed an innovative tied corbel connection detail between the floor steel and the node that permitted the required radial spread of the diagrid during construction while providing a reliable amount of restraint to the diagrid nodes. This detail permitted fine adjustment of the node position during erection, using radial tie bolts. This ensured that the ring of hoop tension elements could be closed without the excessive use of oversized holes or pretensioned bolts. Oversized holes are not generally desired as they can allow elements to be erected out of proper alignment.

The circular geometry also impacted the design of the floors. Radial beams on 10° centerlines span from the steel frame core to the perimeter. This creates a maximum span for the steel deck of 4.75m/15.58ft. The overall depth of the steel pan and concrete topping is 160mm/6.3in, which is higher than normal and provides additional stability to the system.

A view through the vision glass of Swiss Re shows the use of white cladding on the round diagrid members. The "boxes" around the steel are rotated so that they make minimal contact with the curtain wall system. While the diamond-shaped glass panels are fixed, the smaller, triangular windows are operable where required for ventilation. This is critical on the darker colored type of the glass spirals, as these enclose the ventilation shafts.

A view of the interior during construction shows how the metal floor decking spans cleanly between the radial beams, without need for a secondary support system. The radial beams attach to the diagrid.

The triangulation that so dominates this structural solution has required different considerations in the design of the façade. The way that the diagrid is expressed on the exterior of the building is significant in several ways that will be addressed comparatively in *Chapter 9: Façade Design*. But the choices made in the Swiss Re project are worthy of a separate discussion as they served to influence subsequent building designs. As previously mentioned in this section, the diagrid pattern directly influenced the decision to use a curtain wall system composed of a combination of diamonds and triangles to break down the scale of the overall diagrid to fit the desired curved form more closely. The diamond-shaped panes run the full height of the floor with the triangular window elements filling in between and acknowledging the location of the horizontal hoops.

At the base of the building large sections of the diagrid break away from the curtain wall and create a colonnade at the entry points of the tower. Here the cladding of the steel that was used on the interior for fire protection becomes an exterior cladding that assumes the role of weather/corrosion and fire protection. This release of the structure from the building enclosure accentuates the strength of the diagrid as a structural system, in contrast to tall buildings that step back and use an enlarged base for rigidity of the tower/cantilever. Swiss Re actually tapers inward at the base.

The structural systems created for Swiss Re have had a very significant influence on subsequent diagrid buildings, particularly those that have adopted curve-based geometries where triangulation of both the structure and the façade treatment have been critical to the success of the building.

A view from exterior grade up through the triangular shafts. The extreme level of care in the design and fabrication of the multiple cladding systems is evidenced by the coordination of the sizing and joint lines of the systems that converge in this detail.

The diagrid is released from the curtain wall to create a colonnade at the base of the building.

The Hearst Magazine Tower in New York City, NY, USA, designed by Foster + Partners and engineered by WSP Cantor Seinuk, completed 2006, in a view taken from the southeast. The diagrid structure sits atop a historic stone façade creating a remarkable contrast in style, a juxtaposition that drove many design and structural decisions.

A view from the ground floor entrance up into the large atrium space that occupies several of the lower floors of the existing Hearst Magazine Headquarters. Vertical mega-columns and large diagonals are used to support the diagrid portion of the tower that sits atop of this space. The skylight band separates the new and old structures of the building.

PROJECT PROFILE:
HEARST MAGAZINE TOWER, NEW YORK CITY, USA

ARCHITECT
FOSTER + PARTNERS

ASSOCIATE ARCHITECT
ADAMSON ASSOCIATES

DEVELOPMENT MANAGER
TISHMAN SPEYER PROPERTIES

CONSTRUCTION MANAGER
TURNER CONSTRUCTION CO.

STRUCTURAL ENGINEER
WSP CANTOR SEINUK

STEEL ERECTOR
CIVES STEEL

STEEL DETAILER
MOUNTAIN ENTERPRISES

The Hearst Magazine Tower in New York City is perhaps the most normalized of the early diagrid structures, given the rectangular shape of the tower – modified slightly as the corners are indented with "bird's mouths" that serve to accentuate the expression of the diagrid on the building.

The choice and detailing of the diagrid façade were the result of very specific design challenges. Foster + Partners were faced with the requirement to maintain at least the façade of the historic landmark Hearst Magazine Headquarters, while substantially increasing the area of the building. Collaborating this time with WSP Cantor Seinuk as the structural engineers, it was decided to use a diagrid for the structure from the 10th to 44th floor and provide this structure with new foundations that would sit within the floor plan of the existing historic building. This allowed for a separation of both the work and the structure required for the new building from the remediation work for the historic building. The interior of the historic building was demolished and the stone façade was retained and reinforced to the current New York City Seismic Code.

Much of the volume within the existing historic building was given over to a large atrium space, access elevators, services and an auditorium. For structural and aesthetic reasons, the diagrid tower that was to contain the office spaces was stepped back in from the perimeter walls of the existing historic building. This allowed for the construction of a skylight that bridges between the tower and the base building and provides natural light for the lower levels. As the building was looking for (and achieved) LEED™ Gold Certification, natural light was an important consideration in the design.

Access to natural light and a column-free open office plan were also factors that contributed to the choice of the diagrid as the structural system. A diagrid structure could provide sufficient stability on the perimeter without requiring additional columns between the exterior wall and the core. As the site had street exposure on three of the four faces, it was decided to push the core towards the backside of the building, thereby using the floor plan of the building more effectively for light and views. When the core is not at the center of the plan, it loses much of its ability to be one of the primary stability elements for the building. The team looked at varying approaches to the perimeter structure before ultimately deciding on a diagrid solution. It was chosen specifically for its ability to carry both gravity and lateral loads without needing to rely on the core to resist lateral loading. In the end the core was designed as a braced frame steel structure to assist with carrying the lateral loads.

The design of cores for buildings varies geographically in terms of material preferences. New York City has a long history of the fairly exclusive use of steel for core construction, dating back to the time of the Singer (1908), Woolworth (1913) and Empire State Buildings (1930). Only very recently, as a reaction to the aftermath of 9/11, has concrete come into more regular use to provide additional protection to the steel-framed core. This composite system has been used to reinforce the braced core of the lowest 10 floors of the Hearst Magazine Tower in response to the large open spaces and important load transfers between the megacolumns and megabraces and the upper building and its core over the 40ft/12.19m offset created by the large horseshoe-shaped skylight that wraps around three sides of the base of the diagrid portion of the tower.

The diagrid structure wraps around all four sides of the building, thereby creating a perimeter tube. The engineers chose the system for its structural redundancy and the inherent ability of a diagrid to provide multiple load paths. It was also cited by Ahmad Rahimian of WSP Cantor Seinuk as using 20% less steel than a conventional moment frame structure. In looking for LEED™ certification, the steel is noted as having 90% recycled content. The decision was taken very early that the diagrid would not be exposed to view due to a combination of fire engineering and aesthetic choices. Instead, the architect intended to express the diagrid on the exterior and interior of the building through the use of a stainless steel cladding system. This meant that the nodes, and the connections of the diagrid members to the nodes, needed to be very trim in terms of profile size. The intention was that the connection between the nodes and the diagonals be no larger than the outer dimensions of the diagonal members in order to avoid that the structure looks bulky. In the ultimate detailed design of the structure it was conceived as a concealed system.

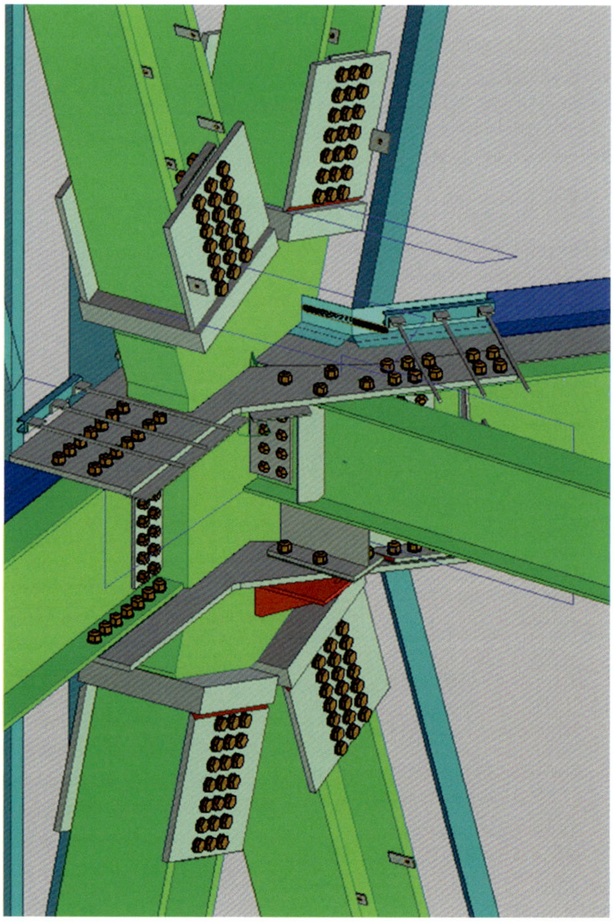

This Xsteel 3D model of a node illustrates the complexity of the bolted connections. The plates that provide the load transfer from the diagrid members through the node are precisely machined to ensure a full load path. Paddle plates are used to allow the bolting to take place at the sides of the diagrid members, thereby keeping the connecting elements "tight" to suit the architectural lines of the stainless steel cladding.

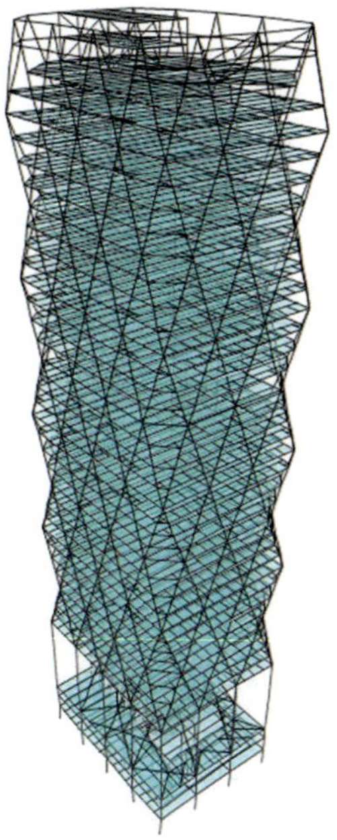

This overall image of the steel structure shows the eccentricity of the core as well as the use of the megacolumns and large diagonal members for the bottom floors of the tower.

Wide flange rolled steel sections (Universal sections) were used for the diagrid members. Their ends were terminated with large plates to transfer the load to similar plates on the arms of the nodes. These were machined to very tight tolerances to ensure good load transfer. The nodes were fabricated from welded plate sections in a basic X-shape to suit the angle of the diagrid while also providing attachment for the horizontal ring beams that act in tension to brace the tower from forces that thrust outward, and providing attachment of a floor beam at the rear of the node via a shear tab. The concealment of the diagrid structure permitted site bolting during the erection of the nodes and diagrid members. It also meant that the choice and treatment of the steel did not need to be AESS driven. This allowed the erection to proceed fairly quickly. For this project the steel erector noted that a combination of standard and oversized holes had been used to allow for some fit adjustment, and that the project proceeded without many alignment or fit issues.

The outer dimensions of the existing historic building are 200ft by 200ft (60.96m x 60.96m) and the dimensions of the diagrid tower are 160ft by 120ft (48.77m by 35.56m). The nodes were placed at 40 ft/12.12m on center to suit the dimensions of the building. Working with this geometry on the façade meant that the nodes would occur in an offset pattern on every fourth floor. As the diagonal members would in this case be spanning four floors between each nodal attachment, the floor beams of the intermediate floors were attached to the diagrid members to provide bracing and reduce their unsupported length. The overall module of the diamond grid is then eight floors, which through today figures on the large end of module dimensions for diagrid structures.

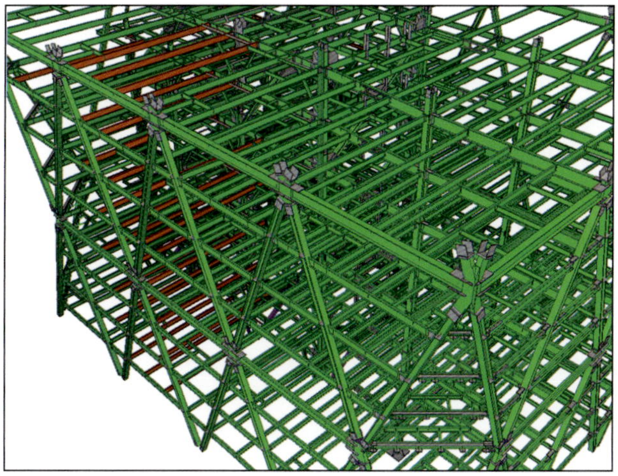

This part model produced by the steel detailer Cives Steel and Mountain Enterprises shows the general layout of the diagrid. The edge beams brace the diagonals between their node connections.

A wood model of one of the corner nodes that was used in the design process.

There are three basic types of nodes that comprise the mid-section of the tower. The nodes at the top and bottom are unique in their configurations as they begin and terminate the structure. The primary node type that is most often replicated is positioned on the wall plane. The nodes at the corners of the full rectangular floor plates were designed to accommodate the specific corner condition, as the incoming members and their load transfer create a situation that is significantly different from the wall condition. The corners of the building have been chamfered back to create "bird's mouths" in order to avoid large cantilevered floor plates between the corner nodes. The nodes at the corner condition have also been refined according to the geometry of the incoming members. Despite these functional reasons for the creation of this detail, it was also recognized by the team as providing a distinct architectural feature in its accentuation of the diagrid structure. The sculpting of the corners by the "bird's mouths" makes the absence of corner columns very apparent.

The development of the structure, connection and details was very reliant on the use of digital modeling. The tight design requirements for the fitting of the exterior curtain wall expression of the diagrid necessitated much exploration of the details, as these had never been developed before. The project team went back to "time-tested" physical wood models to explore the geometries of the nodes, which was reported to have been a highly successful and rewarding process.

Constructability is a critical concern of any steel project and particularly for a tight urban site with little in the way of staging area. New York City, Manhattan in particular, also has issues with the transport of heavy steel sections across the bridges to the island. It can add significantly to the cost of the project if the transportation requires many special permits. If the sections are very large they can only come across certain bridges during the night as not to impede traffic flow. There are weight and size restrictions on all of the bridges that impact the logistics when designing steel structures. Breaking the diagrid into discrete prefabricated nodes and diagonals that could be bolted together in a systemized, highly regular fashion was one of the benefits of the way the diagrid worked for this particular project.

A close detail of one of the "bird's mouths" at the corner of the tower, showing the use of the stainless steel cladding to highlight the diagrid. Fairly standard curtain wall has been used within the triangulation to allow for an easier accommodation of internal shading devices.

Erection, fire protection installation and cladding in progress.

The high degree of repetition in terms of erection sequences and details, as well as the high level of collaboration between the architect, engineer, erector and steel detailer, made the use of the diagrid structure on the Hearst Magazine Tower extremely successful.

The ironworkers from New York City Erectors Local No. 40 receive a node during the erection of the frame. The nodes are basically symmetrical as well as repetitious, so that it was possible to correctly adjust the hook-up to the crane to descend at the right angle to allow for bolting with little force or adjustment to align the members. The connections used a combination of standard and oversized holes to help with site erection.

The cladding and façade treatment for the Hearst Magazine Tower is remarkably different from Foster's previous experience with Swiss Re. Swiss Re uses a triangulated glazing system, appropriate to its curved shape, and keeps the expression of the diagrid's structural members minimal in the cladding. The rectilinear nature of Hearst did not seem to call for the use of triangulated glazing to accommodate the geometry. The large scale of the diagrid module on Hearst allowed for the use of a more standard-looking curtain wall - keeping the triangular or trapezoidal curtain wall fragments to a minimum where the curtain wall meets the diagrid. The trapezoidal elements are also repeated very regularly throughout the façade, allowing for more effective fabrication.

These three initial diagrid projects have formed the basis for the systemization of diagrid structures going forward in the last decade. While London City Hall might have had limited impact due to its scale and partial use of the diagrid, it did begin to address using the diagrid in an architecturally exposed fashion. Swiss Re set forth the language for curved forms that used triangulated façade treatment. Hearst created a larger-scale diagrid form and experimented with an expression on the exterior that included the use of more regular curtain wall infill.

4 TECHNICAL REQUIREMENTS

- DESIGNING FOR PERFORMANCE
- WIND TESTING
- SEISMIC DESIGN
- FIRE PROTECTION SYSTEMS
- —OCCUPANT SAFETY
- —SPRAY-APPLIED FIRE PROTECTION
- —CONCRETE-FILLED STEEL TUBES
- —INTUMESCENT COATINGS

1:50 model created for the shake table test of the Canton Tower in Guangzhou, China, designed by Mark Hemel/Barbara Kuit/IBA Architects with Arup.

Developing a working relationship with the appropriate engineering consultants for wind, seismic and fire-protective design will be an essential aspect associated with managing complex diagrid projects. The purpose of this chapter is to provide an overview of some of the strategies that have been applied to diagrid buildings to date.

DESIGNING FOR PERFORMANCE

Buildings are currently able to push the boundaries on height, geometry and materiality as a direct result of advances in the technology that we now have for design and fabrication. The way that buildings are designed has changed dramatically. Trial and error methods are long outdated, particularly when it comes to structural systems. Former methods of overdesign as a precautionary measure are not acceptable as this adds extra material and weight to a building. Accuracy combined with reasonable factors of safety is the goal of current design. The advances in computing technology over the last 30 years have truly liberated structural design and steel fabrication. Combined with more traditional test methods for wind loading, seismic performance and fire, these are the basis of current engineering practices.

Specific loading issues have a great impact on the design of diagrid structures. One of the most innovative aspects of the way that a diagrid structure works is its ability to resist lateral loads. Traditional moment-framed or braced structures comprised of vertical columns and horizontal floor beams have relied on the lateral stability provided by the core to resist lateral loads. Moving the responsibility of the lateral load resistance to a perimeter diagrid has required the ongoing development of significant research and predictive testing. It is important to realize that existing performance knowledge and more specifically codes are largely based on previous structural types and often do not include diagrids.

The current requirements for wind testing for buildings vary dramatically around the world. In many jurisdictions wind testing is not yet mandatory. It was not until the mid-1960s that wind testing of buildings was even considered as a potential aspect of the design of buildings. At this point such testing would have been considered only for buildings that were very tall or might have unusual snow drifting patterns. This would have been upon the advice of the structural engineer and would not have been mandatory. Major firms whose business focuses on wind tunnel testing only came on the scene in the early 1970s. Yet wind tunnel testing has been around for more than 100 years but was primarily used by the auto and aviation industries to study vehicle performance. The Wright Brothers used a simple wind tunnel to test their airplane as early as 1901.

Seismic loads also have a large impact on the potential lateral load on buildings. Where underperformance of a structure due to excessive wind load might result in occupant discomfort or more minor material or seal failures, seismic design failure has extremely serious consequences. That is not meant to trivialize wind performance requirements but rather to emphasize the difficult task of determining performance under an extreme seismic event. Diagrid structures present a very new and innovative method of structuring a building. With so few in existence there is nothing in the way of a history of performance that can be established via existing structures. Current seismic codes for tall buildings do not reflect diagrid technology.

Fire-protective design is of particular interest because of the architectural implications of this structural system. Where most structural systems tend to be concealed or made to blend in with architectural concepts of a different provenience, diagrid systems are often used as expressive elements *per se*. Fire engineering will determine if the structure can be exposed or must be covered and further protected.

Wind, seismic and fire protection requirements need to be assessed at the very early stages of design as they will determine the range of possibilities for the application of the diagrid to the project. While this is true for all buildings, it takes a special meaning in the case of diagrid buildings as we are only now in the process of establishing successful precedents for diagrid structures in these areas.

WIND TESTING

Wind testing has become an essential part of the engineering of contemporary buildings. Whether based on physical models or digital simulations, wind tests provide information that is critical to designing the structure and cladding and determining the best shape for the building so as not to exacerbate wind conditions in the vicinity of the structure. Often multiple tests will be run to confirm the results of the simulations.

The case of the failed twin towers of the World Trade Center is of significance to the development of high-rise diagrid buildings, as the towers relied on a somewhat similar perimeter steel support system that clear-spanned to a steel-framed core. The National Institute of Standards and Technology for the United States (NIST) study of the World Trade Center disaster cited inconsistency in the predictive wind testing of the failed towers. During its investigation of the collapses, NIST found that wind load estimates from three separate wind tunnel tests on WTC models differed greatly. As a result, NIST has proposed a code change that would require, in the USA, the use of a nationally accepted standard for conducting wind tunnel tests that are used for determining wind loads in the design of tall buildings.

In fairness to the designers of the World Trade Center towers and the engineering firm of Leslie E. Robertson, the use of boundary layer wind tunnel testing for towers was in its exploratory phases at the time. In particular, the traditional all-steel towers in New York City lacked the mass that could be provided by a heavy core, so could not rely on precedents of buildings that used reinforced concrete cores. In order to validate the data being produced in the wind tunnel testing for the World Trade Center towers, they mounted a tower atop the building adjacent to the site and also used two other measuring devices at other locations. The wind speeds were measured and the oscillation for the towers was predicted based on the data collected. It was not known at the time how much movement would be considered tolerable by the occupants. Psychologists were of the opinion that people would simply adapt. The designers were not inclined to take a risk with such a large project, so after developing a number of motion simulators to experience the potential discomfort they decided that a damping system was required. They developed and patented a viscoelastic damping system for use in the buildings to limit the movement.

Regardless of NIST recommendations and spotty adoption into mandatory codes, testing for the effects of wind on buildings is now routinely carried out for all tall buildings and any building with unusual geometry that would adversely affect human comfort or result in excessive snow loading. Human comfort would include the impact of the wind pattern on the environment at grade as well as motion in the upper levels. For tall buildings, wind loads must be known in order to properly design the cladding system and attachments. There are three ways that wind tests are executed:

→ Computational Fluid Dynamics (CFD)
→ Boundary layer wind tunnel
→ Water flume

The majority of buildings will use Computational Fluid Dynamics (CFD) as well as one or more of the physical tests. Lower-rise buildings will be tested in a water flume to assess for accumulations of drifting snow or sand. One of the key considerations in determining the structure and shape of a diagrid tower will be its ability to resist wind loading. Although CFD programs are able to do some predictive modeling that will help to determine form, most very tall or unusually shaped buildings are tested in a physical wind tunnel.

The wind engineers will suggest changes to the shape of the building based upon their findings. The investigation will also look into the design of any damping systems that will be required to offset the potential sway at the top of the tower due to the effects of wind. The Tuned Mass Damping (TMD) systems must be accommodated into the plan and section of the building. Extremely tall structures tend to need to be designed more aerodynamically.

Testing is of high importance for those new diagrid structures that are comprised of twisted or unusual shapes as the engineering profession does not have any standard "rules of thumb" on which to rely. In the case of Swiss Re (30 St. Mary Axe) the form of the building was developed in response to the wind patterns around the site to create a friendly pedestrian precinct at the base of the tower.

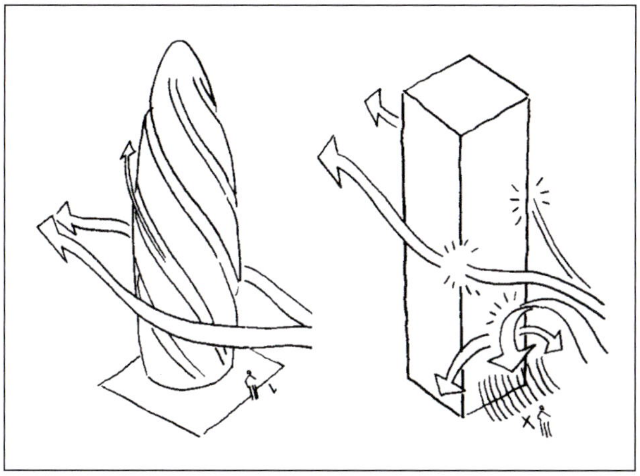

The form of the Swiss Re Tower (also called 30 St. Mary Axe) in London, England, designed by Foster + Partners and engineered by Arup, was intended to capture the flow of the wind around the building to assist with the natural ventilation strategies created by the spiral double façade atrium elements. Foster's sketch compares his tower with the standard rectangular tower that does not allow the wind to flow around the building and therefore makes a harsh wind regime at the base of the building.

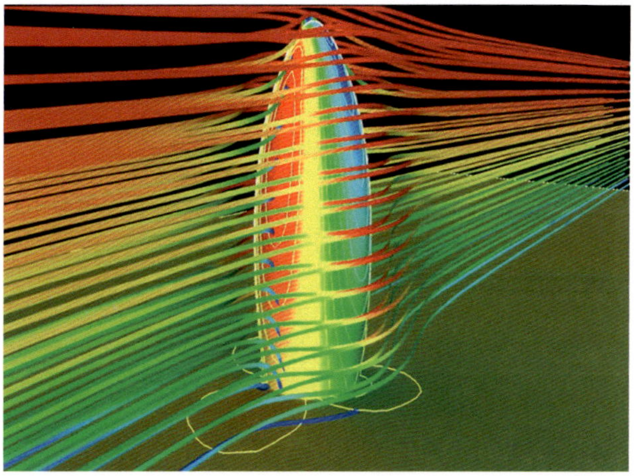

The wind flow patterns around Swiss Re as projected due to its form.

Where the sketches and CFD model may be useful in establishing the initial understanding of the actions of the wind on the building, these sorts of drawings and models do not address the major issue of the context of the building. Urban situations, in particular, have extremely complicated wind flow patterns as a direct result of the configuration of buildings adjacent to the subject site being studied. It is most efficient to do physical wind tunnel modeling to address this issue. This is normally done in a boundary layer wind tunnel.

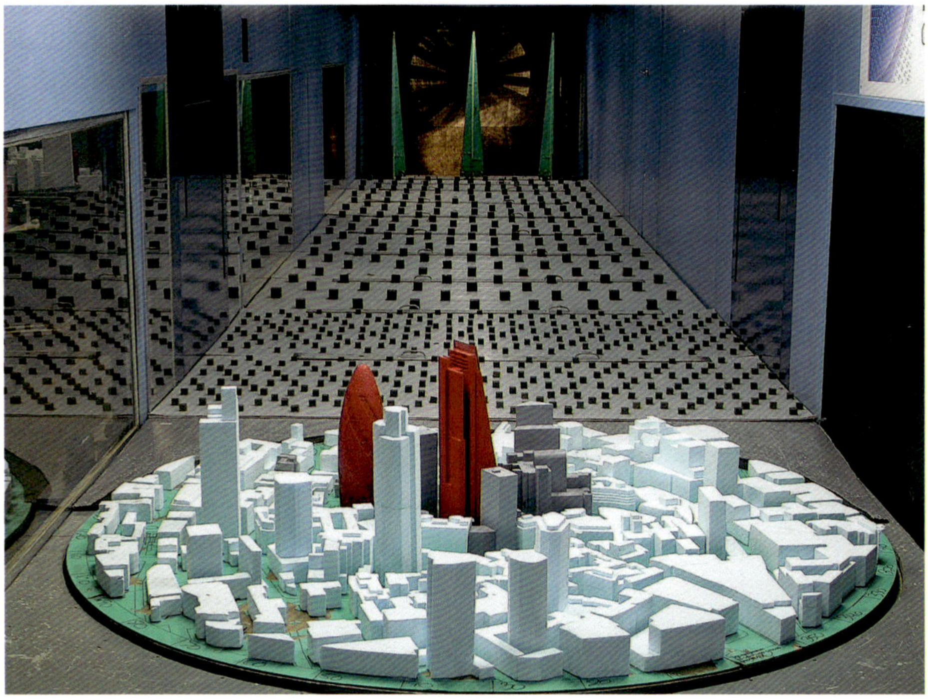

When buildings are placed in the wind tunnel it is important that the terrain around the building be modeled as well as its immediate urban environment (boundary layer) in order to provide an accurate simulation of the resultant wind pressures. This is the model for The Leadenhall Building in London, England, designed by Rogers Stirk Harbour + Partners with engineering by Arup, as tested in the facilities of RWDI. You can recognize Swiss Re and the very specific form of the towers adjacent to the building. This specificity is important to produce accurate results.

The types of physical models that are used in the boundary layer wind tunnel are executed in great detail. It is important to model the texture of the cladding and the precise geometry of corner conditions as these will affect the readings. This is easily done using 3D printing technologies. Models can be created at different scales to assess different aspects of the design. Smaller-scale models will be used to look at the urban situation in general as can be seen in the image of the boundary layer model done for The Leadenhall Building. Buildings that are situated in a more isolated situation are sometimes modeled larger. The size of the model will be determined by the capabilities of the wind tunnel. Commercial wind tunnels range in cross section from 2.4m x 2.4m/8ft x 8ft to 9m x 9m/27ft x 27ft. The standard boundary layer wind tunnel will have a rotating table to allow for ease of study of the effects of the wind from various orientations.

Varying the architectural form away from rectilinear can be used to "fool the wind". This is scientifically termed "vortex shedding" where uneven geometries are used to prevent the vortexes on the leeward side of the structure from aligning, which if in tune with the natural resonance of the building, could cause catastrophic failure. Much study is being done on the designs of the corners of buildings as these have a significant impact on vortex shedding and the way that wind flows around a structure.

It is very important to assess the effect of the wind at the base of the building, particularly in the instance of large towers in urban settings. Funneling of the wind in dense urban sites can make the pedestrian precinct uninhabitable. A closer view of the wind model for The Leadenhall Building fitted out for the wind test for the pedestrian precinct: the model at left shows the pre-construction condition and at right the constructed condition.

The Canton Tower in Guangzhou needed to respond to a number of critical lateral loading issues. The 600m/1,969ft communication tower is located in a typhoon-prone area. Although the occupants would be evacuated during a typhoon event, the deflections and movements of the tower still need to be limited. Elliptical plan-form shapes like that used for the Canton Tower are prone to across-wind excitations caused by vortex shedding. It was necessary to determine the accurate wind load and to set the correct lateral drift deflection and building acceleration limit. The tower's unique diagrid structural form was not covered in any building codes regarding its determination of wind loads, which was one of the critical load cases to be studied. Arup wind engineers led a series of wind climate studies and wind tunnel laboratory tests to accurately assess wind loading. They used the Monte-Carlo simulation technique to deduce the wind profile, a problem solving technique used to approximate the probability of certain outcomes by running multiple trial runs or simulations using random variables. These included a topography test and large-scale 1:150 sectional wind model test. A numerical method was used to integrate wind forces into further structural design. The large-scale sectional model was also used to test the micro-climate behavior of the tower's open zones as the external diagrid structure type had also no wind precedent, particularly at this large scale.

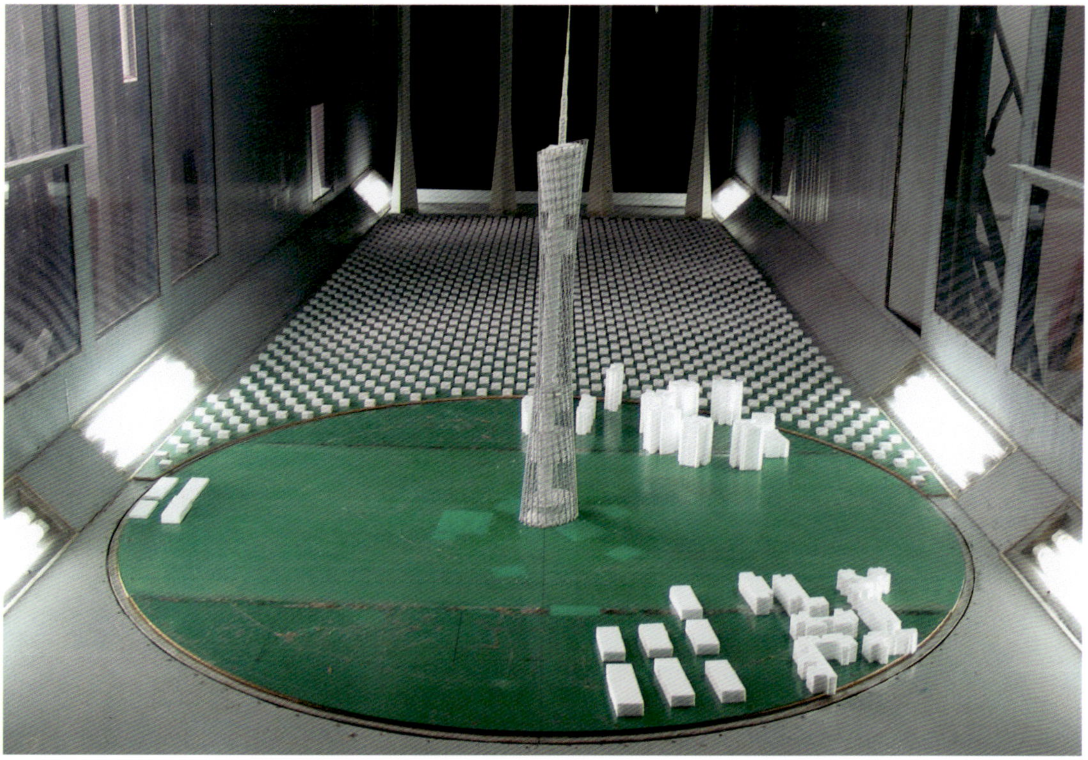

A view of the Canton Tower during the boundary layer wind tunnel test that was directed by Arup. The absence of other buildings adjacent to the tower reflects the actual site conditions as the tower is situated by the river and there are no other tall structures nearby. It can be seen that the mid-height of the tower consists only of the exterior diagrid and the elevator core, making the study of its strength extremely critical at this point.

As the Canton Tower is a broadcast tower it was important that this function could continue during high wind events. Additional testing was carried out on the antenna portion of the tower to ensure that movement would be limited. A damping system in the top of the tower and in the mast section was developed to assist in controlling movement.

More information on wind testing can be found in *Understanding Steel Design, Chapter 9: Advanced Framing Systems: Diagrids.*

SEISMIC DESIGN

The establishment of seismic performance factors for steel diagrid-framed seismic-force-resisting systems is a relatively new area of engineering study. Current model building codes do not explicitly address diagrid systems. Under moderate to extreme earthquake ground-shaking demands a typical seismic-force-resisting system needs to be able to provide sufficient ductility and energy dissipation characteristics. This is required in order to prevent collapse while undergoing inelastic frame deformations. In the normal triangulated configuration of a perimeter diagrid-framed system, the gravity and lateral loads are distributed through the sloped column and spandrel beam (horizontal tie) members as axial tension or compression. Under the axial compression conditions the diagrid-frame elements will be designed to remain linear elastic with an appropriate factor of safety.

If the steel perimeter diagrid system is constructed with a reinforced concrete core, it is considered by seismic design to be a dual system – the reinforced concrete core considered to act in a ductile fashion. If the diagrid building is designed without a reinforced concrete core, it cannot act as a dual system and lacks the ductility provided by the core. In this case, the perimeter diagrid system becomes a standalone seismic-force-resisting system and requires a completely different method of design and analysis. This can be termed a "bearing wall system" or an "unidentified system" according to some code provisions. It is necessary for undefined systems to be tested in other ways to determine performance.

Where standard computational methods may not be sufficient to predict the performance of a diagrid structure, physical shake tables can be used. Shake table testing was used on the Canton Tower as its diagrid form lay beyond standard seismic design analysis. One of the challenges of structural stability was the narrowing of the tower at its "waist" and the relative lack of structural material at that section available to resist loads. Advanced computational techniques were used to verify that the tower would withstand severe seismic events as well as check the slender column length at mid-section, where there is no floor diaphragm. Large-scale model tests including a 1:50 shaking table test and a 1:15 stability load test on the waist were carried out to verify structural safety.

The interior structure of Manukau Institute of Technology, Auckland, New Zealand, designed by Warren and Mahoney Architects, uses a seismic bracing system to reinforce the structure. The diagrid in this case is supporting the perimeter loads but is not the sole structural steel system for the building. The deep floor plates and interior courtyard are structured in a fairly standard way to comply with the New Zealand codes. The Eccentrically Braced Frame (EBF) was fabricated for MJH Engineering by D&H Steel from 60mm/2.4in flange plates with a 50mm/2in web using full penetration welds.

The diagrid buildings constructed to date, particularly the towers, have been erected in areas of low to moderate seismicity. None have been constructed in highly seismic zones, with the exception of the Manukau Institute of Technology in Auckland, New Zealand. This low-rise building only uses a diagrid to support two façades. The steel in the other portion of the building uses seismic bracing techniques via the use of Eccentrically Braced Frames (EBF).

As at the time of writing there is not a defined code-based approach for assessing the seismic performance of a diagrid structure. It is anticipated that research will continue in this important area. In the interim, testing approaches for signature projects will continue to use a combination of methods of analysis including physical modeling and shake table testing.

The shake table test of a section of the Canton Tower. The tower was outfitted with a Hybrid Mass Damping system: this is comprised of a Tuned Mass Damping (TMD) system with two stage damping levels, and a compact Active Mass Damping System (AMD) which is driven by linear induction motors mounted on the TMD. In case of a failure in the HMD control system the system would become a passive TMD.[1]

A large-scale model of the Guangzhou IFC in Guangzhou, China, designed by Wilkinson Eyre Architects with Arup. The innovative nature of the application of these large diagrid structures necessitates methods of exploration and structural validation that often include the construction of large physical models.

FIRE PROTECTION SYSTEMS

Although steel is an incombustible material, it is subject to sudden collapse when exposed to heat during a fire. While the specifics vary by jurisdiction (use, building height, and access to fire-fighting equipment), steel is generally considered to need fire protection if expected to endure a fire for more than 45 minutes. In most fire codes, the focus of fire-protective design is firstly to maintain the structural integrity of the building long enough to evacuate the occupants and secondly for the structure to remain intact after the fire has been extinguished. The larger and more complex the building, the more difficult this becomes. This means that the use of simple exposed painted structural steel is quite limited. Fire protection can be achieved either by covering the steel with a material that will delay the heat damage for a period of time, or through the use of a fire suppression system, or with a combination of both.

For a more complete discussion of the various options for the fire protection of steel it is recommended that you refer to *Understanding Steel Design, Chapter 7: Coatings, Finishes and Fire Protection.* The contents of this chapter will only address specific innovations found in diagrid buildings.

Occupant Safety

Issues surrounding the safe evacuation of the occupants may lead to differentiated requirements for the protection of the steel structure during a fire event. On very tall buildings the use of evacuation floors is now commonplace. These floors are provided with increased fire resistance and are used to stage the safe evacuation of the occupants so that the normal means of egress are not overburdened. The use of evacuation-designated floors was part of the strategy used on the Guangzhou IFC. The instances of the floors are expressed in the banding on the façade of the building.

The façade treatment on the Guangzhou IFC creates a design element from the evacuation and mechanical floors. The evacuation floors are placed where required for fire engineering purposes and are structurally identical with the other floors of the building. In an outrigger system, the reinforcing outrigger floors perform a double duty: they are evacuation floors and must simultaneously consider structural and fire requirements to determine their location.

Spray-applied Fire Protection

Spray-applied fibrous fire protection is commonly used in concealed structural steel framing systems. This method has replaced the use of sprayed asbestos fireproofing. There are three different compounds used in this method: gypsum plasters, cementitious plasters and fibrous plasters. The materials are spray-applied to a requisite thickness in order to delay the rise of the temperature of the steel to its approximate failure temperature of 540°C.

The Al Bahar Towers in Abu Dhabi, UAE, designed by Aedas Architects with Arup, used a spray-applied fireproofing on the structural steel members, as this was designed as a concealed steel system and would be clad.

The gypsum-clad diagrid members use a faceted hexagonal cladding to minimize the impact of the wide flange sections on the interior space. The exterior mullion expression of the diagrid on the north face can be seen on the exterior of the curtain wall.

The floor structure of Capital Gate in Abu Dhabi, UAE, designed by RMJM, uses a spray-applied type of fire protection as the elements will be concealed. This system is used in conjunction with an intumescent system for the AESS perimeter diagrid structure. Mixed protection systems are often used to suit finish and exposure issues.

Concrete-filled Steel Tubes

Concrete has long been used as both a fire protection method as well as a composite material for structural reinforcement of the building. Concrete can be either used to surround the steel, creating a depth of cover between the fire source and the steel, or to fill HSS sections, thereby increasing their resistance to fire. Filling provides less resistance than encasement and depending on the intended time of fire resistance required may also need additional protection. The use of concrete-filled tubes is seeing increased use in Asia and the Middle East as a strategy for providing fire protection in composite construction. It is also seeing increased adoption as the method of choice for tall diagrid structures seeking to expose their structural systems.

A concrete-filled steel tube column system has many advantages compared with standard structural steel or plain reinforced concrete systems when constructing large diagrid structures.

- Better interaction between steel tube and concrete: Local buckling of the steel tube is delayed and the strength deterioration is delayed due to the restraining effect of the concrete. The strength of the concrete is increased due to the confining effect provided by the steel tube. Concrete spalling is prevented by the tube. Creep of the concrete is smaller than in ordinary reinforced concrete.
- Cross-sectional properties: The steel ratio in the concrete-filled tube cross section is much larger than in reinforced concrete and concrete-encased steel cross sections.
- Construction efficiency: The labor for forms and reinforcing bars is omitted and the concrete casting is done by Tremie tube or pump-up methods. This efficiency leads to a cleaner construction site and a reduction in labor, construction cost and project duration.
- Fire resistance: Concrete improves the overall fire resistance so that fireproof material such as additional intumescent coatings can be reduced or omitted.
- Cost performance: Because of the merits listed above, better cost performance is obtained by replacing an all-steel structure with a concrete-filled tube structure.
- Environmental impact: The environmental impact can be reduced by omitting the formwork, specifying steel with a high recycled content and using high-quality concrete with recycled aggregates.

There is a certain elegance in the round and smooth forms of concrete-filled steel tubes that is not easily replicated by other types of steel sections, with or without a fire resistance covering or casing. It is only suited to members with large cross sections.

These very large tubular columns at the Guangzhou IFC are filled with concrete as part of their fire resistance and have a trowel-applied coating to meet the required rating for this class of Supertall tower.

A view inside one of the concrete-filled steel tubes for the Canton Tower. These are all custom-fabricated. The ring stiffeners on the interior provide added strength at points of load transfer.

The preference for the use of large steel tubes with concrete fill is regional. This type of construction is very common in China but is seldom seen in Europe or North America. The Western markets tend to make use of hot-rolled sections or built-up plate when larger sections are required. While this construction method would have been expected on the Guangzhou IFC and Canton Tower, it is interesting to see that the structural diagrid system used for the Doha Tower in Qatar, designed by Ateliers Jean Nouvel, uses a similar concrete-filled tube system. The detailing of the diagrid of the Doha Tower is very similar to the one used on the Guangzhou IFC. Given that the engineers and contracting company used for the Doha Tower were Chinese, a globalization of construction methods may show itself here.

The all-welded HSS diagrid columns that create the structural and focal point for the new entry pavilion at the World Financial Center in New York City, designed by Pelli Clarke Pelli, use intumescent fire proofing to allow for this exposure of the structure. The finish works well with the uplighting system. The form is very reminiscent of the original hyperbolic paraboloid diagrid towers designed by Vladimir Shukhov. Steel fabrication was done by Walters Inc. and erection by Metropolitan Walters.

Differences in labor rates and construction safety concerns also tend to drive the choice of structural type. Great concerns over worker safety in North America, Australasia and Europe tend to aim to diminish the amount of welding done on site. There is a very high amount of welding required on the connections for concrete-filled tubular steel systems, which involves a high amount of work being done at height.

Intumescent Coatings

Intumescent coatings provide both fire resistance and a painted appearance for exposed steel. They contain a resin system that is pigmented with various intumescent ingredients which, under the influence of heat, react together to produce an insulating foam or "char". This char layer has low thermal conductivity and extends to a volume many times that of the original coating. The char layer reduces the rate of heating experienced by the steel, thereby extending its structural capacity. As this material can extend the fire resistance rating of exposed steel to a maximum of two hours, it has become quite popular for AESS applications in diagrid structures. The fire resistance rating is in part dependent on the type and thickness of the coating as well as the type of fire that might be anticipated in the building use. Increasing the fire resistance rating is usually achieved by applying multiple coatings of the product. The coating will have a slightly thickened appearance with a finish akin to an orange peel. This needs to be taken into account when deciding on the details for the connections and members.

Intumescent fire protection is used on the architecturally exposed steel structure of Capital Gate. The expression of the members and their connections is a very important aspect of the design and so an intumescent coating was the applicable choice.

The exposed structure of The Leadenhall Building made use of intumescent fire protection to permit its structure to be expressed as part of its double façade system.

5

MODULES AND MODULARITY

The expression of the diagrid module on the Aldar Headquarters in Abu Dhabi, UAE, designed by MZ Architects, highlights the way in which the module and modularity of the diagrid structural system informs the architectural expression of the building.

ISSUES OF SCALE AND SHAPE

The challenges associated with designing with diagrids vary greatly with the type of building, its geometry and size, and much of this is reflected in the choice of the module size for the building. This chapter will examine the relationship between the size of the module, the physical arrangement of the building, efficiency and form. This will build upon current optimization research that is based on quantitative investigations.

Both the Guangzhou Opera House by Zaha Hadid and the Guangzhou IFC by Wilkinson Eyre Architects use diagrid structures. The decisions surrounding the structural design of these two buildings could not be more different in spite of their common preference for a modular, triangulated steel structural system.

The diagrid shape and module size for a building must satisfy the structural requirements as well as the aesthetic intentions of the project. Scale is a critical issue. If we think back to the initial use of expressed diagonals on tall diagonalized core towers such as the John Hancock Center and the Bank of China Tower, the diagonals were extremely large in cross section and created a distinct, grand pattern on the façade.The massiveness of the members at their point of intersection with the vertical columns was capable of obliterating significant views from the adjacent floor areas; however, these interruptions were infrequent due to the overall scale of the diagonal bracing system. A diagonalized core tower often highlights the expression of the diagonals on the exterior while attempting to minimize the impact on the interior by taking advantage of the immense size of the structural members. Such a strategy is conceivable thanks to the ability of the diagonal bracing to form a very clear visual strategy on the exterior of the building while making rather sporadic appearance in the interior. There are simply not enough diagonal members to make a significant contribution to the design of the interior spaces.

The large bracing members in KK100 in Shenzhen, China, designed by TFP Architects, (top left) are framed out in gypsum board. They have an infrequent appearance in the space so do not form a dominant pattern in terms of the interior design or space planning.

The large diagonal braces on the John Hancock Center (top right) block off significant portions of glazing.

The large diagrid members at the hotel lobby level (70th floor) of the Guangzhou IFC (left) have been integrated into the décor. Their frequency, as a function of the chosen module, allows them to become highlighted in the interior layout.

As diagrid buildings are often aesthetically driven, the diagrid is usually highlighted on both the exterior and interior appearance. The scale of the diagrid members, their physical shape and treatment, and frequency can be used to enhance the experience of the interior spaces. Regardless of fire engineering decisions that will affect the ability to expose the steel members or require them to be clad, the diagonals are usually part of the design aesthetic in the interior.

STRUCTURAL PERFORMANCE CRITERIA

A diagrid is a structural system that gains its stability from triangulation. As with triangulated trusses it is assumed that the nodes act as hinge or pin connections and are not moment-resisting. This would infer that the loads acting on the diagrid members are axial (tension or compression) and that only shear forces are transferred through the nodes. This may be true for extremely simplified analysis, but practically speaking, to achieve constructability, there needs to be adequate stiffness in the connection between the node and diagrid member to be self-supporting in order to minimize the need for temporary bracing during erection. This is less of a concern for low-rise structures as it is less difficult to provide temporary shoring for a low building than it is for a very tall building in the case that the node connections were to be designed with less stiffness. This will be discussed further in *Chapter 6: Node and Member Design* and *Chapter 8: Constructability*. Another factor is that the actual loading on the diagonal members will change as a function of the form of the building. As was discussed on Swiss Re (see page 39), the inward or outward lean of the perimeter tube changed the forces in the "hoops", or horizontal bracing rings, from tension to compression.

The structural performance criteria for a diagrid system will be quite different for low-rise versus high-rise structures. High-rise structures will have much higher lateral loads due to the actions of wind and potential seismic incidents. The diagrid tube for a tall building must act like a vertical cantilever and be very stiff. Generally speaking the diagrid system for a tall building will need to resist moment forces at the base and shear forces toward the top. Low-rise buildings that use a diagrid system to support floor loads will not have the same severity of moment loads at the base conditions. So for the design of the module for a tall building, the structural requirements suggest that it vary along the height of the building to suit the different lateral loading. Module design must also take into consideration the natural increase in member size to accommodate the accumulation of gravity loads on the lower floors of the structure. The 103-floor-tall Guangzhou IFC uses 2m/6.56ft diameter concrete-filled steel tubes for the lower floors with 1.1m/3.61ft concrete-filled tubes for the upper floors to reflect the extreme differences in loading.

MODULE SELECTION CRITERIA

One of the initial tasks that must be addressed by the collaborative team is the selection of the module of the diagrid:

- How frequent are the nodes?
- How long are the diagonals?
- What sort of spacing angle is appropriate?
- What is the spacing between the points of connection with the floor edge beam that creates the triangulation between the diagrid members?

The selection of the module and the general modularity of the diagrid greatly differ for low-rise versus high-rise applications. Within these two types, the level of consistency of the geometry will impact the design of the diagrid and the size of the module. Modules will be different for angular versus curved applications.

The following considerations will impact the design of the diagrid system and selection of the module:

- geometry of the building
- occurrence of eccentric loading
- structural efficiency
- floor-to-floor heights
- requirements for fenestration pattern and window sizes
- selection of AESS or concealed steel structure

There are additional criteria that impact the module size for tower type buildings:

- height/width and proportion of the building
- core design (ability or not to assist with lateral load resistance)
- wind and seismic loads

On a tower, the module refers to the number of floors measured vertically between connecting nodes, tip-to-tip of the diamond shape.

The size and arrangement of the modules impact the structural design, member sizing and node design for the building. These choices, in turn, have a direct impact on the architectural design.

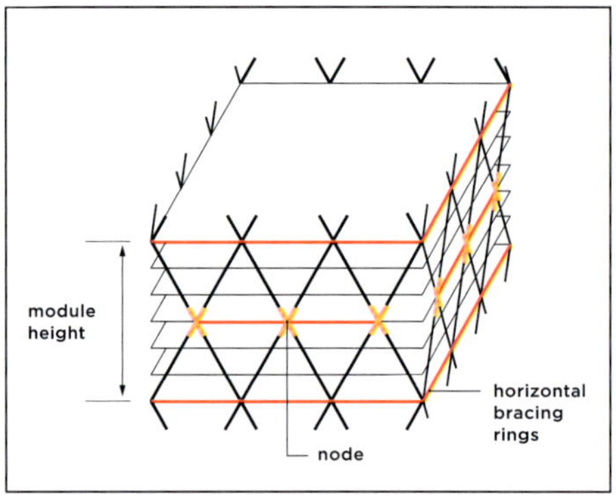

Basic terminology for structural diagrids as applied to regular geometries, typically for tower structures of varying height. The module height for this example would be six storeys.

Many diagrid buildings, such as the Addition to the Royal Ontario Museum in Toronto, ON, Canada, designed by Daniel Libeskind, use non-regular modules. Here the faces of the "Crystal" are subdivided to suit the geometry. The spans and the roofing or wall-support structure is sized to suit the resulting spans.

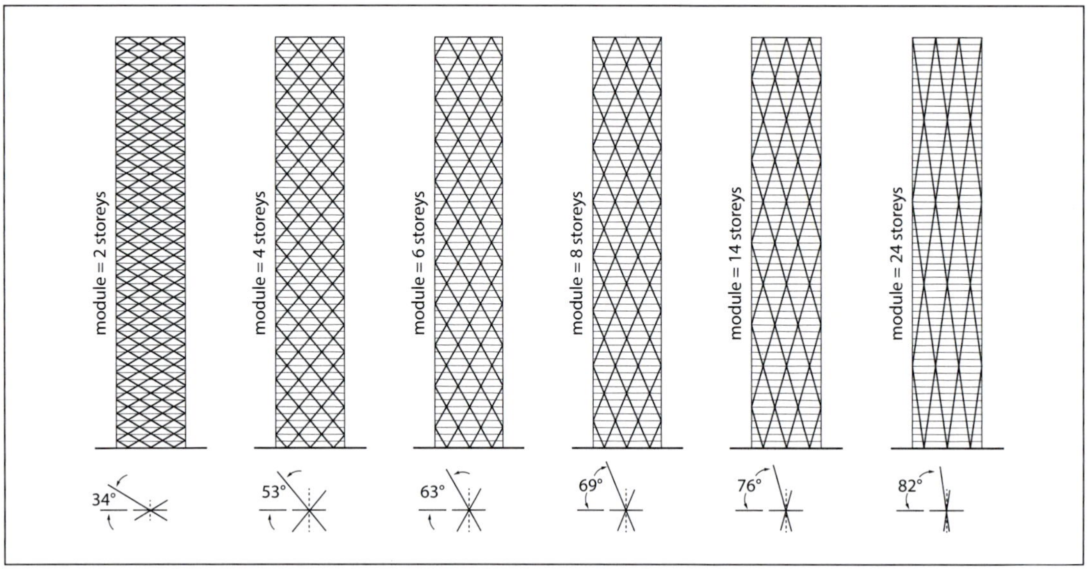

This image, modeled after the diagrid research work of Kyoung Sun Moon, shows the visual impact of the change in the inclination of the diagonals on this sample 60-storey tower. All sample schemes maintain the same plan dimensions in order to minimize impact of the variation on the design and span considerations of the floor structure.

All of this assumes that the building itself has some inherent modularity and is making use of a regular module. However, diagrids are increasingly being used for buildings with irregular geometry, as the diagrid is an excellent way to respond to providing a stability system for such projects. In these cases, the sizing of the modules is in principle derived as a subdivision of the geometric shape.

OPTIMIZING THE MODULE FOR STRUCTURAL PERFORMANCE OF TALL BUILDINGS

Much engineering research is underway to establish the optimal module size, which directly impacts the shape of the diagrid, the window size and placement as well as the amount of resources used in the project. The impact of shape is viewed differently by architects and engineers - structural concerns also being impacted by wind and vortex shedding issues. The primary university-based research has been undertaken by Kyoung Sun Moon of Yale University with studies on building heights of 40, 60 and 80 floors, and more recently studies on diagrids for twisted towers. The sample study done of a 60-storey-tall building was based on a 1:6 ratio, measuring 36m x 36m (118ft x 118ft) with an 18m x 18m (59ft x 59ft) gravity core at the center and floor-to-floor heights of 3.9m (12.8ft). This study confirmed 69° as the most effective angle for a uniform diagrid as it results in the least amount of steel by weight. This angle changes as a function of the building height as well as the building's width-to-height ratio.

Moon's studies found that for diagrid structures with uniform angles – i.e. assuming a square plan that is consistent throughout the height of the building – the optimal angle that the diagonal makes with the floor increases as the building becomes taller. Taller buildings with a large height-to-width aspect ratio tend to behave more like a bending beam. Steeper-angled diagonals can resist these bending moments more efficiently. For tall diagrid structures with height-to-width aspect ratios ranging from 4:1 to 9:1, the range of the optimal angle is from 60° to 70°.

Given that a tall building functions as a vertical cantilever, the differentiation between the function of the diagrid elements at the base to those toward the top will change the taller and more slender the building is. This has suggested that a variation in the inclination angle of the diagonals can be used to reduce the amount of steel and contribute to the required resistance in the structure. Generally speaking, the members and connections at the base of the building must be designed to resist moment or bending, while those at the top must be designed to resist shear.

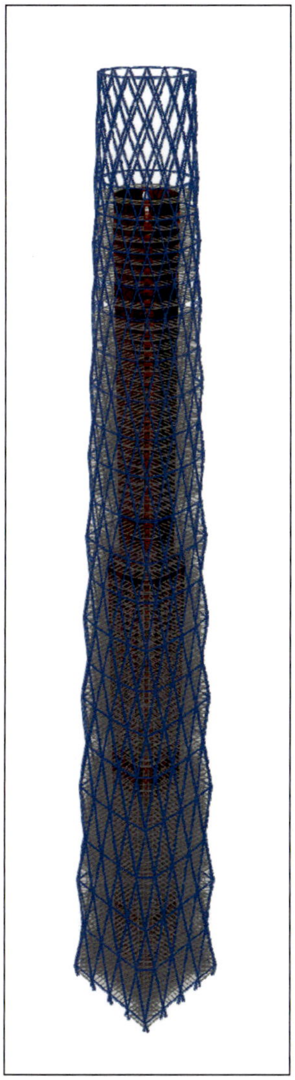

The modules and angles for the diagrid on the proposal for the Lotte Super Tower in Seoul, South Korea, designed by SOM, use a variation in the angle and module height to respond to the changing lateral forces on the tower. The angles are steeper at the base and more acute at the top, as the gravity loads are more significant at the base and the wind loads more severe toward the top of this 555m/1,821ft-tall tower.

Referencing the figure illustrating the range of modules from 2 to 24 (see opposite page), it must be understood that the diagrid members from projects with smaller modules will be easier to transport and handle than the diagonals on projects with larger modules. This infers the need to accommodate splices in the diagonal members for large modules. The impact of making splice connections on the diagrid members will be different for exposed versus concealed systems. The choice of member and connection type will greatly impact the cost and scheduling of the project. Very long members might be able to be spliced prior to lifting, but there might be weight or balance issues associated with such heavy members that preclude such a strategy. This is where good communication among the team members is essential in order to understand the complete ramifications of intermediate splices on very long diagrid members as a result of large modules. This will be discussed in more detail in *Chapter 8: Constructability*.

Moon has also conducted studies on towers where the angle of the diagrid varies along the height of the tower in acknowledgement of precisely this change in structural requirement from moment at the base to shear toward the top. For aspect ratios of 7:1 or more it was found that less steel than for a uniform angle was required where the angles toward the base of the building were steeper and became shallower toward the top. For buildings with a height-to-width aspect ratio less than 7:1 it is the uniform angles that produce a more efficient design. The reduction in the amount of steel will translate into cost savings on the steel tonnage as well as economical modifications in the foundation design. Naturally, it must be appreciated that these findings are very general and that the specific requirements will vary according to the engineering design of each project.

The design process for the Lotte Super Tower in Seoul, designed by SOM, explored a significant variation in the use of modules. The tapering and transforming geometry of the building, combined with the optimized diagrid structure, would have allowed the unbuilt proposal of the Lotte Super Tower to be constructed with approximately 27% less steel than a conventional steel-framed tower, confirming Moon's studies. The columns toward the base have an angle of close to 80° in order to better resist the accumulated gravity loads. The angles would gradually change toward the top, where the angle decreases to close to 50° to better resist the high wind loads. Additionally, the plan changes from a 70m/230ft square base to a 39m/128ft circle at the top, to address wind loads and vortex shedding. For additional information on the design of the Lotte Super Tower please refer to the *Project Profile* on page 172.

BRACING OF THE DIAGONAL MEMBERS

The triangulation of the diagrid "tube" itself is not sufficient to achieve full rigidity in the structure. Ring beams at the floor edges are normally tied into the diagrid to integrate the structural action into a coherent tube and connect the tube to the floors, and back to the core. As there are multiple floors intersecting with each long diagonal member of the grid for modules of four storeys and greater, these intersections will occur both at the nodes and at one or several other instances along the diagonal. This reduces the unsupported length of the diagonal member, providing restraint and preventing buckling. The angle of the diagonals allows for a natural and direct flow of loads through the structure and down to the foundation of the building.

The bracing strategies for the long diagonal diagrid members will impact the overall design detailing of the structure. In some cases the horizontal ring bracing beams or hoops will be made evident in the façade, resulting in the triangulated appearance of the curtain wall. In other cases the floor edge beams are more uniformly tied into the diagonal members and the horizontal ring beam will tend to recede in appearance. These latter types of buildings will sometimes tend to express more of a diamond-shaped pattern on the façade (see *Chapter 9: Façade Design*).

The Bow Encana Tower in Calgary, AB, Canada, designed by Foster + Partners with Zeidler Partnership, uses a 12-storey module. The diagonal members have been fabricated from plate steel to create a triangular cross section. As evidenced by the white-colored areas surrounding the nodes, where site welding has occurred, the six-storey-tall diagonal members were able to be transported in one piece. This greatly simplified the site work. In this project the diagonals are not braced by the floors between the nodes, because of the double façade atrium space behind the diagrid.

In the Swiss Re Tower (30 St. Mary Axe) in London, England, designed by Foster + Partners with Arup, the different functions of the members are reflected in the structural sections. The diagonal diagrid members are fabricated from round tubular steel as these need to be heavier in order to resist compression and tension. The diagonals span two storeys node-to-node so that some are braced by the floors at the mid-point. The diagonals that cross the ventilation shaft are unbraced yet have been kept to the same exterior dimensions as the braced members in order to permit the use of the same details for the cladding and create a more uniform set of connections to the node. The "hoops" that create the tension bracing system around the circumference of the tower are formed by connecting the more slender edge pieces to the nodes. These are connected in a different manner than the primary diagonals to the node.

The floor edge beams on the Aldar Headquarters are evenly tied into the diagonals to provide a fully braced scenario. The edge condition of the floors is consistent throughout the façade and allows for a diamond-shaped cladding pattern that does not accentuate the horizontal connections between the nodes.

The edge beams of the Guangzhou IFC are all tied into the diagonal members, creating a fully braced system.
This was necessary because of the extreme height of the building, the long member lengths and the moderate seismic and high wind loads.

The construction sequence for the Aldar Headquarters uses a fast track scenario. The erection of the steel diagrid lags slightly behind the pouring of the concrete core. Glazing modules are mounted over the diagrid that has already received its spray fire protection (white). The expression of the diamond modules is clearly evident throughout the construction of the building. Steel fabrication and erection by William Hare Limited.

In their quite similar overall structural design, the Guangzhou IFC and the Doha Tower in Qatar, which use concrete-filled steel tubes in the creation of the diagrid, both rely on the edge beams of the floor system to fully brace the long diagonal members.

In a drive toward structural efficiency in the Lotte Super Tower in Seoul, the diagonals are rigidly connected to major horizontal spandrel beams on multi-floor modules occurring at the intersection of adjacent diagonals. This allows the spandrel beams to participate in the lateral system by transferring the lateral loads between the diagrid and the floor slab.

MODULES AND CORNER CONDITIONS

A key aspect of the diagrid is its continuity as it wraps around sharp changes in the plan shape of the building. The larger idea of the diagrid as a perimeter tube system is the notion of the continuity of the tube, in combination with the elimination of vertical columns. While this is fairly simple to achieve on a building with a "rounded" plan, it is more difficult on a plan with sharp corners. If we refer to the diagram of a rectangular prototypical building (see page 64), it can be seen that where the diagrid wraps the corner condition no structural support is provided for the floor slabs. The cantilevered slab condition can present structural issues on buildings with large modules or where the horizontal spacing of the diagrid nodes is wide. Smaller buildings with more closely-spaced nodes and tighter modules will present less of a challenge in supporting the cantilevered corner condition, as distances will be significantly smaller.

In the diagrid buildings featuring a rectangular plan that have been constructed to date, the challenges presented by the corner condition have been used to enhance the visual strength of the design. It should be noted that for very tall buildings, there are additional concerns in the design of corners as they impact the action of the wind on the building including vortex shedding. Corner detailing will require the input and expertise of a wind engineer.

Warren and Mahoney Architects of Auckland, New Zealand, refer to the perimeter diagrid that completes two sides of the Manukau Institute of Technology building in this way:

"An exposed raking column steel frame [diagrid] on the north and east elevations is the key feature of the exterior appearance of the building and required close integration of the architectural and structural design. Project Director Jeremy Austin for Holmes Consulting Group explains: 'Because of their geometry and their position outboard of the main structure, the raking frames bear high axial and torsional loads. The raking frame members are welded box sections connected by three types of nodes: Y-nodes at the base, X-nodes at mid-height, and inverted V-nodes at the top. Since the nodes are exposed in the finished façade, their design had to be simple and neatly detailed. Full-penetration welds of the 40mm/1.6in plate steel were needed to ensure adequate transfer of the forces. All welds were ground flush to provide a high level of finish. The connections are fully bolted site connections; site welding was avoided for durability and aesthetic quality reasons. Full tension friction bolts were secured to captured nuts inside the box sections'."

At the Manukau Institute of Technology in Auckland, New Zealand, designed by Warren and Mahoney Architects, the strategy strikes a compromise between the discontinuous support of the floors by the diagrid and the expression of the structure. In this instance, a lighter steel system of a contrasting color is set in behind the primary diagrid to physically support the floors with a discreet corner column.

The Hearst Magazine Tower in New York City, NY, USA, designed by Foster + Partners and engineered by WSP Cantor Seinuk, has solved the issue of the cantilevered corners by removing them. The exterior lines of the building follow the diagrid, creating a "bird's mouth" feature that runs up the corners of the building. This adds stability to the diagrid frame that might otherwise have been compromised by the large cantilevers, which would have been 20ft/6.1m long at their greatest distance.

The corner design for One Shelley Street in Sydney, Australia, designed by Fitzpatrick + Partners with Arup, has allowed the floor edge beams to cantilever from their last contact with the diagrid. The diagrid members that wrap around the corner to complete the exterior structural frame stand out free of contact with the building structure.

IMPACT OF THE MODULE ON THE NODE

When considering the efficiency of a diagrid structure, it is important to take a very holistic view of the project. Diagrid buildings such as the Hearst Magazine Tower and the Bow Encana Tower have been reported to use 20% less structural steel than a conventional moment-framed building. While this is important in terms of project and environmental costs, it must be recognized that diagrids increase the fabrication costs due to the use of specialized connections - the nodes - even though these are normally prefabricated. From the point of view of fabrication and cost, the module size and consistency has a direct impact on the design and fabrication of the nodes. Larger modules require fewer nodes. As the fabrication costs of the nodes are likely to greatly exceed the base cost of the steel, larger modules can offer fabrication and erection savings.

Buildings that are more regular in their geometry will require fewer types of nodes. A good case for this would be the Hearst Magazine Tower, which is rectangular in plan and consistent along the height of the building. Although different nodes are always required at the base and top of building conditions, the majority of the conditions are addressed through the design of the basic types of wall and corner nodes.

A round tower that has a consistent plan dimension throughout its height would maintain the basic geometry of the node for the full building - naturally accounting for a change in the size of the diagonal members and loading requirements that are always larger at the base of the building than toward the top. Consistency does simplify the design strategy for the nodes in terms of their appearance, connections to the diagonals and fabrication processes.

Towers whose plan dimensions vary over the height of the building will require modified nodes for each floor level. This was the case in Swiss Re. The gherkin shape of the tower also impacted the design of the hoops at the floors: the forces changed from tensile in the lower two thirds of the tower to compressive toward the top, as the diagonals start to lean in toward the core. However, the regular geometry of the module design allowed for the nodes to maintain a consistent appearance and fabrication methodology.

SIZES OF MODULES FOR BUILDINGS OF DIFFERENT SHAPE AND HEIGHT

The application of the modules to various sizes and shapes of buildings falls into relatively distinct categories as pertains to the overall height of the module. The height of the module will have a great impact on the ability of the floors to be used for bracing the diagonals. The length of the diagonals will impact fabrication, transportation and erection. As will be further explored in *Chapter 9: Façade Design*, the size of the module pattern will have inferences on the selection of the glazing system and pattern.

Small Modules: Two to Four Storeys

Smaller-sized modules of two to four storeys of height tend to be applied either to buildings with a small height or those with very unusual geometries or load eccentricities.

The module used on Capital Gate in Abu Dhabi, UAE, designed by RMJM, is one of the smallest to date. The diamond-shaped module measures two floors from tip-to-tip. This infers the absence of an additional support along the diagonal members at mid-height. The reason for this tightly knitted structure is the accommodation of the stresses that arise from the 18° backwards lean of the tower. The lateral loading is only partly taken by the diagrid structure. A prestressed, "pre-cambered" concrete core has been used to balance the eccentric loading. Although the angles that the diagrid members make with the floors vary throughout the entire building, they hover around 45°, which results in a relatively "square" diamond pattern. As a result of the twisted form of the building, all 8,250 diagrid members are different in thicknesses, length and orientation and each of the 822 diagrid nodes is unique. The module size for the interior atrium is consistent with the exterior module. However, the selection and detailing of the members and nodes are very different.

A module of four floors from tip to tip of the diagrid is used on the Swiss Re Tower. In this instance, the relatively small module is required in order to allow for the use of straight diagrid members to achieve the appearance of a curved shape without necessitating the bending of the members. The tighter module also contributes to the intention to spiral the darker blue ventilation shafts up the volume of the building. The changing building shape results in an increasingly steeper grid angle toward the top of the building.

The diagrid at the Manukau Institute of Technology is quite unusual for being installed external to the envelope as well as having the nodes occur at a mid-floor height. This makes for a module of five storeys from tip to tip. The small scale of the institutional building uses a steep angle that one would expect on a taller structure in order to make its architectural statement. Instead of a horizontal ring beam at the centerline of nodes, the structure uses a tension system to tie the diagrids together. The diagonals are braced at the two points where they pass by the floor edges.

The Tornado Tower in Doha, Qatar, uses a four-storey module on its façade. The module size allows the tower to alter its shape quite easily as it adapts to the hyperbolic paraboloid shape, reminiscent of some of Shukhov's original tower designs.

The diagrid for One Shelley Street in Sydney, Australia, designed by Fitzpatrick + Partners, is quite unusual in its exterior use of the structural grid. The module measures only two floors from tip to tip and the connection back to the floors and ring beam occurs only at the nodes in order to limit the penetrations through the curtain wall. This negates the possibility of bracing the diagonals along their length. The angle on the node is the least steep of any constructed thus far, being 38° from the horizontal. The wrap of the exterior diagrid around the corners is also highly unusual. The low angle flattens the diamond modules, accentuating the width of the blocks rather than their height.

Midrange Modules: Six to Eight Storeys

The middle range of module size is suitable for larger buildings and those whose geometry is more uniform. As the length of the diagrid member will range from three to four storeys, transportation and erection is usually not a great issue. This length is often assembled in the staging area, with the node to create an assembly in the shape of an inverted V for erection. This minimizes the connection work of the ironworkers high on the building, saving time and cost.

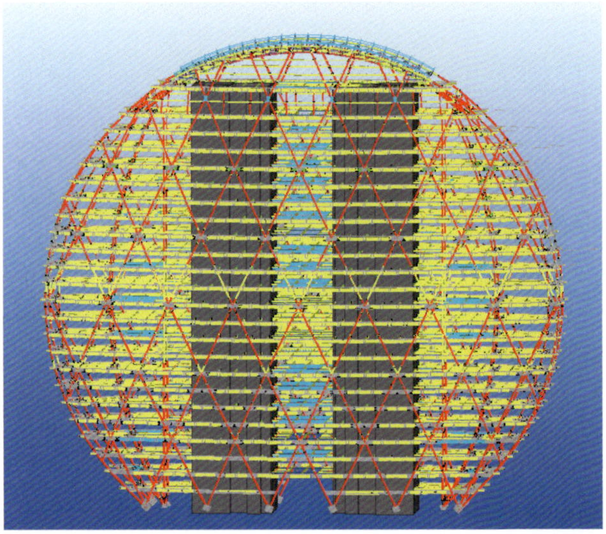

An eight-storey module has been used on the Aldar Headquarters. Although this is a comparatively large module for a 23-storey building, and the angle that the diagonals make with the floor is steep, the very thin floor plates minimize the load on the diagrid structure. The horizontal ring beams that occur at the level of the floor slabs have been downplayed in the façade, leaving a diamond pattern.

The Hearst Magazine Tower uses a six-storey module. The angle that the diagonals make with the horizontals is 69.7°. This creates a visual verticality to the diagrid that accentuates the height of the building. The exterior diagrid wall has been extended above the roof level to hide the mechanical equipment.

The Doha Tower in Doha, Qatar, designed by Ateliers Jean Nouvel, uses a mashrabiya screen to clad the exterior of the tower. The eight-storey module is visible through the façade screen only at night. The Tornado Tower beyond (blue lighting) uses a smaller, four-storey module. Many towers integrate lighting to highlight the diagrid expression at night.

Large Modules: 10+ Storeys

Modules that measure 10 storeys or higher are only appropriate for use on very tall buildings out of considerations of scale. As the support points for the connection of the nodes to the floor edge beams (horizontal rings) cannot be excessively large, having to relate to the spanning capabilities of the floor beam and slab system, this naturally creates a diagrid module that is proportionately much taller than it is wide. The height of tall buildings is thereby accentuated, making them look even more slender than they may be. The overall length of the diagonal members will require special coordination regarding shipping-related issues. It is always desirable to minimize on-site welded connections; if these members are oversized due to their module length, one may incur extra costs on site for assembling.

The use of concrete-filled steel tubes that is becoming common in China and parts of the Middle East will always require significant site welding as the connections must be sealed. This system is becoming a common choice for large module diagrids constructed in these regions as can be seen in the Guangzhou IFC and the Doha Tower.

The modules used on the Guangzhou IFC measure 12 storeys on the lower three quarters of the tower and 16 storeys at the top. The lower portion is used as office space, with floor-to-floor heights different from the top hotel floors. The module height is consistently 54m/177ft. The long diagonals are fully braced by each floor slab, allowing downplay of the triangulation that is normally created by horizontal bracing rings that occur between the nodes at the floor levels. The combination of the curved form and narrowing of the tower towards the top result in slightly steeper diagrid angles at the base (more advantageous for moment resistance) and slightly more inclined angles toward the top (more advantageous for shear resistance).

The Bow Encana Tower uses a 12-storey module for its south façade. As the diagrid members on this face are separated from the building floors by the empty space inherent in a double façade atrium, they had to be extremely large because no bracing was possible. The diagonal length was shipped as one piece.

The selection of the module on The Leadenhall Building in London, England, designed by Rogers Stirk Harbour + Partners with engineering by Arup, had to work with the "cheesegrater" shape of the building that was created to maintain the required visual sightlines to St. Paul's Cathedral. The 14-storey module has been incorporated into the design, based on its seven-storey coordination with the overall building form. The diagonals are braced at each floor level. The ring beam between the nodes has been brought forward and accentuated as it plays a role in the design of the double façade system.

Irregular Modules

Buildings that are highly angular will use modules that are irregular and sometimes chaotic in their appearance. The module size will be carefully mapped to the desired shape of the structure and will need to respond to quite different criteria as a direct result of the unique nature of the application of diagrid framing to this sort of structure.

In some instances, where the diagrid is designed to provide structural support for façades that are not intended to be fully glazed, the modularity (or lack of it) of the diagrid tends not to be expressed in the cladding. In these cases, the subdivisions of the modules are based more on spanning distances and wind deflection criteria of the secondary support systems, rather than on an architectural desire to express the modularity or pattern of the diagrid in the façade treatment.

A diagrid structure was used to support the crystalline shapes in the design for the Addition to the Royal Ontario Museum in Toronto, ON, Canada by Daniel Libeskind. The diagrid is highly irregular with no two members or connections being identical. A project like this relied heavily on the BIM-based detailing software to manage the complexity. The expression of the diagrid on the exterior and interior of the building is quite different from the more regular tower type applications.

The module used on the CCTV Building in Beijing, China, designed by OMA with engineering by Arup, is very irregular. The pattern created by the acknowledgement of the module on the façade relates to the placement of the diagrid according to calculated positions of stress. The vertical elements have been suppressed to further highlight the diagonal pattern. The very eccentric loading of the building and the decision not to use the core to assist with lateral load resistance put all of the loading requirements on the primary steel structure.

Irregular modules tend to present even more challenges when it comes to the structural design and planning for the fabrication and erection sequences for the project. The lack of modularity will create many unique geometries and circumstances. The use of digital technologies has become even more critical to support this type of steel construction. Prior to the advent of this type of software this level of complexity would not have been considered feasible.

A diagrid structure has been used on the Canadian Museum for Human Rights (CMHR) in Winnipeg, MN, Canada, designed by Antoine Predock. The structure on the left side of the building will include some exposed steel and is highly glazed. The structure on the right uses a diagrid approach to support the angular shapes and is clad in stone. The module for the clad section always spans the same number of floors but uses different angles to work with the geometric changes in the tower shape.

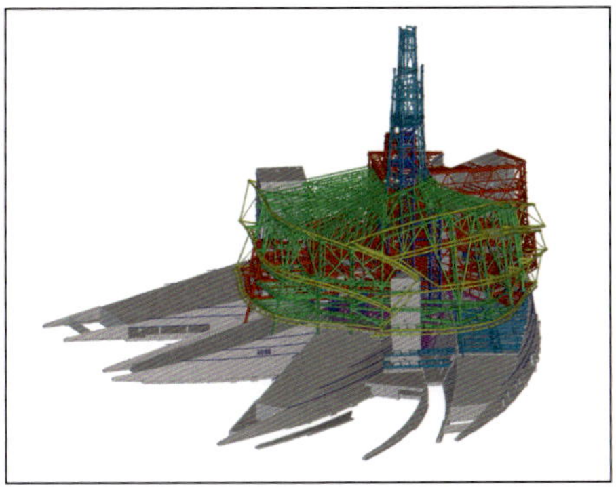

The use of a BIM model for the structural design of chaotic steel geometries is essential. This image shows the Tekla model of the CMHR, prepared by the steel fabricator Walters Inc. The steel framing towards the rear, depicted in red, was designed as a concealed steel diagrid that supports stone cladding.

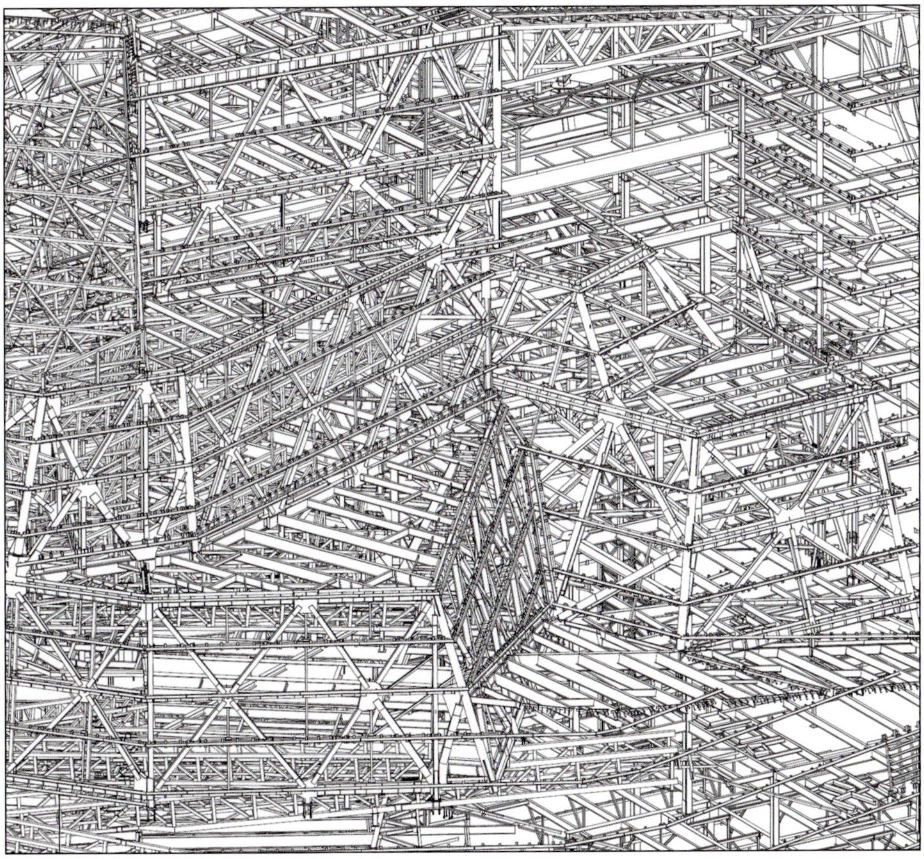

This close view of the 3D structural steel model for the Canadian Museum of Human Rights by the fabricator, Walters Inc., shows the intensity and density of the steel structure, revealing the diagrid elements as well as the many additional steel members required to frame such a complex structure.

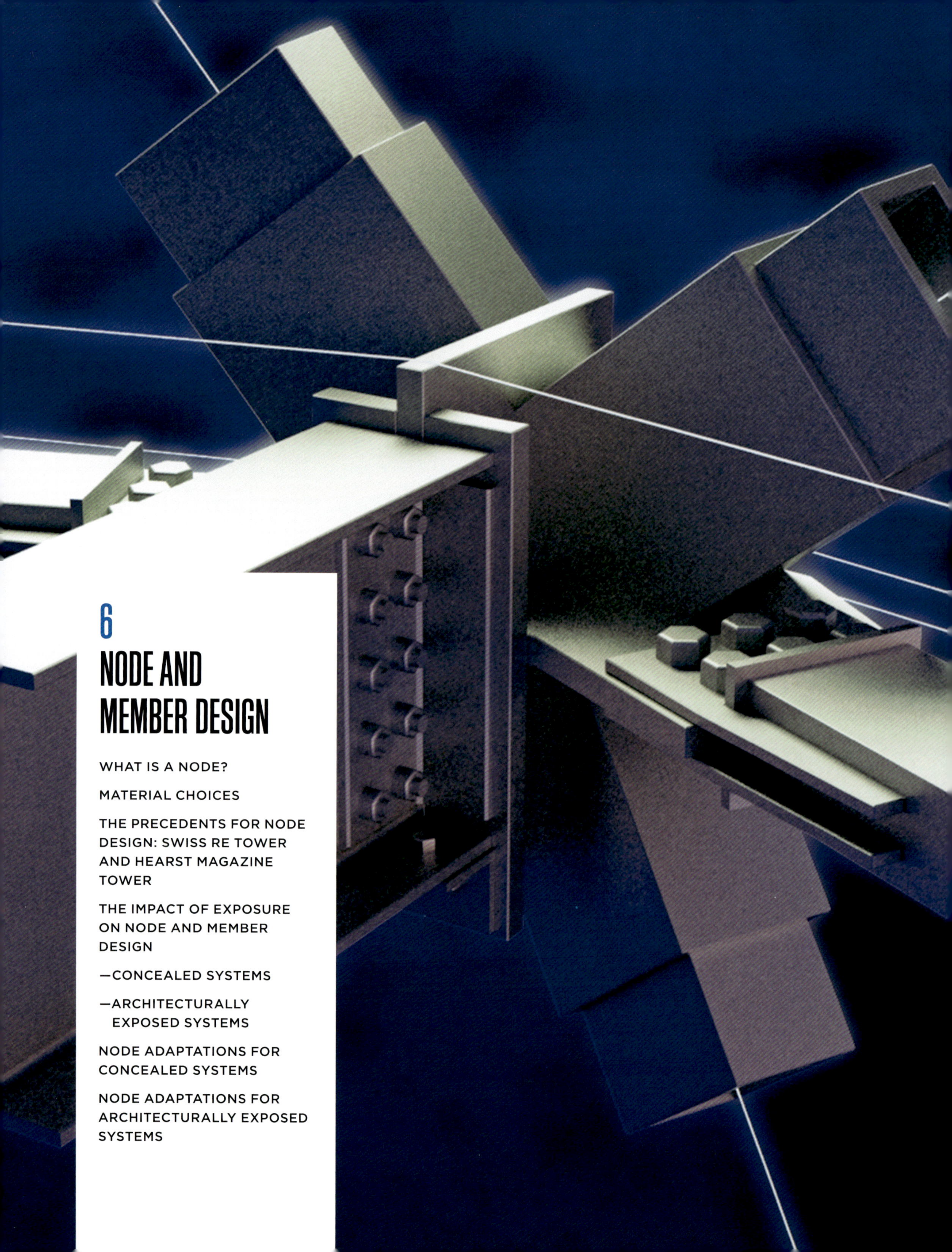

6

NODE AND MEMBER DESIGN

WHAT IS A NODE?

MATERIAL CHOICES

THE PRECEDENTS FOR NODE DESIGN: SWISS RE TOWER AND HEARST MAGAZINE TOWER

THE IMPACT OF EXPOSURE ON NODE AND MEMBER DESIGN

—CONCEALED SYSTEMS

—ARCHITECTURALLY EXPOSED SYSTEMS

NODE ADAPTATIONS FOR CONCEALED SYSTEMS

NODE ADAPTATIONS FOR ARCHITECTURALLY EXPOSED SYSTEMS

Three-dimensional models were used to explore the configurations of the nodes for Capital Gate in Abu Dhabi, UAE, designed by RMJM. Jeff Schofield, one of the designers for the project, created numerous iterations to evolve the design. Even through to the final constructed building the nodes can be seen to change as a result of collaboration with the fabricators. The connections to the incoming diagrid members were eventually changed to be parallel to the ground rather than in line with the X-shape of the node.

WHAT IS A NODE?

A diagrid structure is constructed by connecting the diagrid members to a series of nodes in a determined, typically triangulated pattern that is composed of modules. Thus far we have looked at the relationship of the modules to the shape, size and structural requirements of the building. The selection of the module size and angle infers requirements of the nodes and diagrid members.

It is really the scale, member size and node design that differentiates a perimeter-support structural diagrid system (supporting floor loads) from a lattice grid system. While a lattice grid system is designed to support glazing and opaque elements, the drastic increase in the loading requirements for floor support increases the structural size of the diagrid members and therefore the nodes that they connect to. Lattice grid systems take their design and assembly cues from three-dimensional spaceframe systems, flattening out the detailing to become more planar.

The lattice grid system that supports the curved glazed wall at the Ottawa Congress Center in Ottawa, ON, Canada, designed by Brisbin Brook Beynon Architects, is constructed of custom-fabricated structural steel designed to standards of Architecturally Exposed Structural Steel (AESS). In this type of structure, columns are used to support the floor loads. The rectangular HSS members that form the lattice grid are joined by custom-fabricated nodes that stack in a split fashion to allow for easier assembly. The scale and erection techniques for this type of system are very different from a structural diagrid.

Lattice grids must be designed to accommodate the direct attachment of glazing systems and therefore will have smaller tolerance levels. However, the triangulated nature of the lattice grid makes it quite suited for use in conjunction with diagrid systems, particularly where the module size of the diagrid is large and the fenestration pattern is triangular. Structural lattices can effectively be used to subdivide the diagrid and provide good support for glazing systems. This will be discussed further in *Chapter 9: Façade Design*.

MATERIAL CHOICES

Most nodes for diagrid systems tend to be fabricated from custom-cut welded plate material. This material seems well suited to crafting the highly articulated shapes that are capable of accepting four to eight incoming members. The way that the steel can be cut will influence the level of detail as well as the amount of remediation required. Materials of differing thicknesses use different processes. These can result in a rough or smooth cut edge. Modern steel-cutting equipment can work with Computer Numerical Control (CNC) systems for great accuracy. Manual cutting requires more clean-up depending on the skill of the operator and the applied AESS level (see *Understanding Steel Design, Chapter 6*). The options of cutting technologies are limited by the thickness of the steel plate.

Methods include:

- Plasma cutting: The thickness of steel with this method is typically ¼in to 1¼in / 6mm to 30mm.
- Oxy-fuel cutting: This is the most common method and the thickness of material is unlimited.
- Water-jet cutting: This method is less common and the limits on steel thickness are not known.
- Laser cutting: This method is used on material in the range of 1/16in up to a practical limit of ¾in / 1.5mm to 20mm.

Although castings have evolved as an alternative means of solving the load transfer for complex connections, they have not been adopted into diagrid structures to the time of writing. More information on castings can be found in *Understanding Steel Design, Chapter 10: Castings*.

THE PRECEDENTS FOR NODE DESIGN: SWISS RE TOWER AND HEARST MAGAZINE TOWER

The Swiss Re Tower and the Hearst Magazine Tower provided the precedents for subsequent designs of diagrid structural systems. Swiss Re tackled curves and used round tubular structural members. Hearst took on a rectangular plan and used hot-rolled wide flange or Universal sections. Although neither system was designed using Architecturally Exposed Structural Steel (AESS), the aesthetic demands of the cladding systems required that the connection designs be very "tight" in order to diminish the bulk of the clad members.

When these projects were in the development and detailing stage, Building Information Modeling (BIM)-based steel detailing software such as Xsteel (now Tekla Structures) was available, if not as advanced and interoperable as it is today. A prerequisite for both projects was the need to shop-fabricate the nodes and diagrid members so that they could be more easily erected on site. The connections within the nodes themselves were fully welded. All of the on-site connections were achieved through bolting. Shop fabrication of connections allows the effective use of jigs to ensure that the alignment and positioning of the elements is accurate. The shop is also equipped with cranes and other innovative devices that can lift, turn and rotate the members so that there is easier access for welding and remediation operations such as grinding and milling, resulting in a higher quality of the final product. When the steel of the diagrid system is to be exposed, the quality control of the shop environment for welding is essential.

The prefabricated node for the Swiss Re Tower (30 St. Mary Axe) in London, England, designed by Foster + Partners and engineered by Arup, provides four identical attaching points for the incoming diagrid members and an alternate system to connect the bracing hoops, which in this case are acting in tension.

The prefabricated node for the Hearst Magazine Tower in New York City, NY, USA, designed by Foster + Partners and engineered by WSP Cantor Seinuk, was designed with bolted plate connections to accept the incoming diagrid members. As can be seen in this image, this node is situated at one of the corners, with the plate attachment for the incoming horizontal bracing system set at angles to suit the geometry at the corner condition.

What becomes evident when examining the nodes for Swiss Re and Hearst is the importance of ensuring that the forces are axially transferred through the nodes. The alignment is most evident on the Swiss Re node as one can see the plates converge at the center of the node. Although the node with the connections to the diagrid members is stiff for purposes of constructability, the element is not intended to be moment-resisting beyond assisting with the erection processes and the force transfer should be similar to that of a truss-type design.

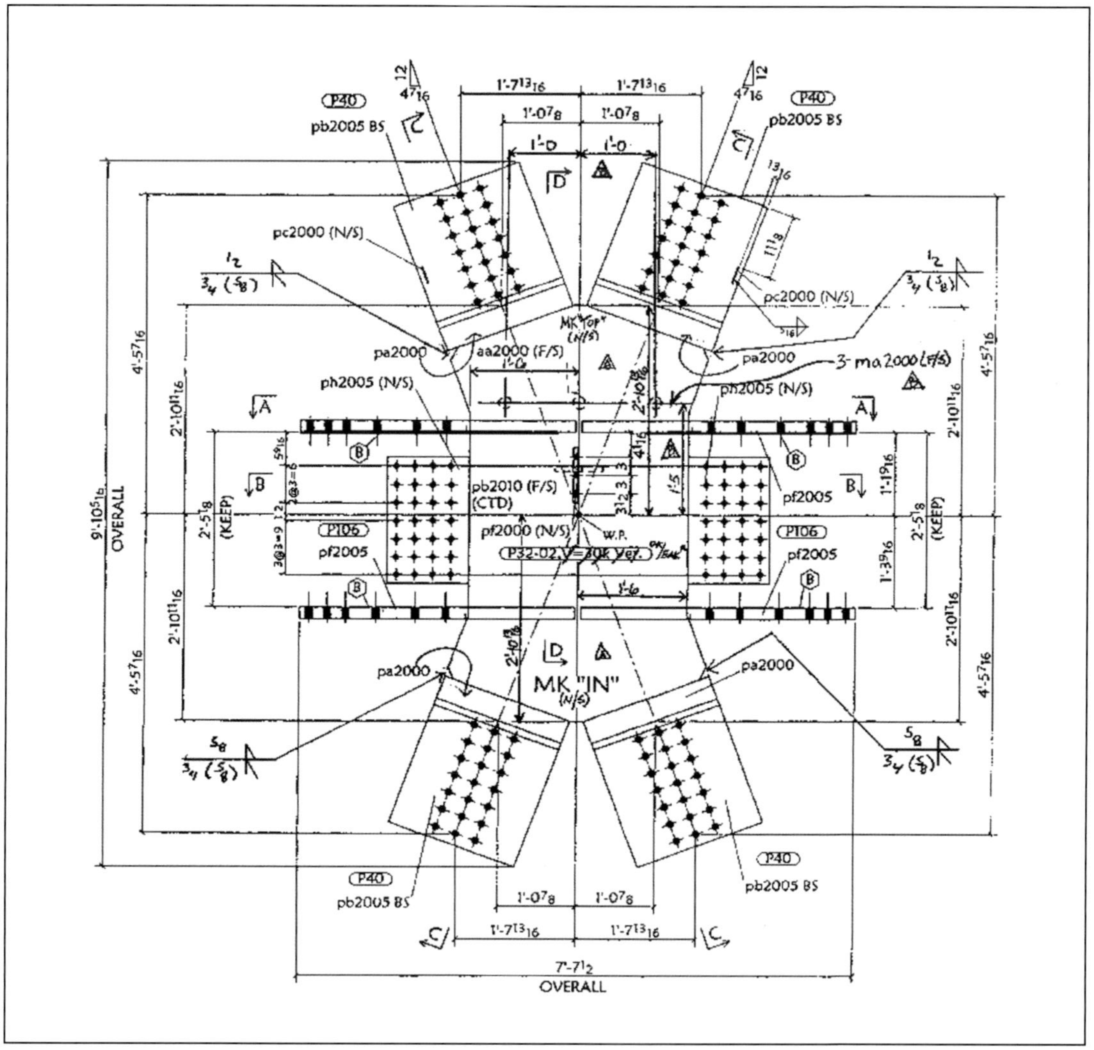

This detail drawing of a typical node on the Hearst Magazine Tower illustrates the fairly standard way of setting out the angles of the diagrid structure. The structure was detailed by WSP Cantor Seinuk using BIM software.

The nodes of Swiss Re and Hearst both use very thick plates for the transfer of load from the diagrid members to the node. At Swiss Re the plates visibly match in size and are bolted together, with the tubular diagrid members having round plates welded to their end points. It was noted as very critical for achieving a uniform load transfer that these plates be machined to tight tolerances. On the Hearst node, the main load transfer plate has been attached to the node, but there is no matching plate on the diagrid member. The plate on the node and the end of the wide flange member are milled for thorough contact and the force transfer occurs partly through this plate and partly via tabbed connecting plates welded to the thick plate on the node at right angles. These connecting plates bolt to plates that were shop-welded to the flanges of the wide flange members. It was important for the cladding design that these side plates be tight to the profile of the wide flange beam.

THE IMPACT OF EXPOSURE ON NODE AND MEMBER DESIGN

The choice to architecturally expose or conceal the structural diagrid frame has great impact on the design and detailing choices for both the member types and nodes. There may be instances where the structural system needs to be less visually dominant as a function of the use of the space; however, whether or not a diagrid is exposed or concealed, there will always be a strong tendency of the diagrid to dominate the design. Exposure is not always an option, as fire codes vary by building type and jurisdiction, so it may not be permitted to make the expression of the diagrid in this way an integral part of the design. A proper fire engineering analysis must be carried out on the project to ascertain the eligibility of the structure for exposure. The type of fire protection system for the structure will greatly impact the cost of the structure. The type of coating system, resultant finish and the protection of that system from damage will be critical to consider when detailing the steel.

Capital Gate (right) is able to expose its powerful diagrid, making the steel a vital part of the interior expression of the building. The Royal Ontario Museum (left) must clad the members as part of the fire protection system. The gypsum board cladding changes the feeling of the materiality of the steel in the space. With clad structures the work and presence of the structural system seems less apparent. However, with diagrid structures, even clad diagonals will have a greater design impact than vertical load-bearing systems that can be hidden or not expressed.

Concealed Systems

The primary concerns that will influence concealed steel diagrids are loading and erection issues. Complete shop fabrication of the nodes has become the industry standard. As these elements are large and normally welded, overhead cranes are needed to move and rotate the steel for ease of access for welding. It will be more likely to use bolting to join the diagonals to the nodes, as this can be done more quickly on the site. However, this does not preclude welding the nodes to diagonal connections on site. This is the standard for concrete-filled tubes. In preparation for on-site welding the node and diagonals will arrive with connection tabs attached. The tabs are bolted during assembly to temporarily secure the steel so that the crane can be disconnected. The tabs will be cut away once the welding is complete. In the case of concealed systems the resulting surface imperfections will not need to be remediated.

Instead of concrete-filled tubes, concealed structural steel diagrids more often use basic types such as wide flange or Universal sections for the diagrid members. These are fairly simple to work with and the connection of the member to the node can be handled with bolted plates. Swiss Re is one of the few examples to date to use tubular sections for concealed diagrid members; this was done in part to make the profile work with the circular geometry of the building. Where the loads are very high, custom-welded plate sections have been used. The exact choice of the member will vary with the building type and loading conditions.

In cases where the diagrid is chosen purely for its structural efficiency and is not expressed, or at least largely hidden under gypsum board, the detailing will be less of a concern for the architect. However, if the diagrid will be clad as part of the architectural expression (see for instance Shelley Street, p. 71), then the connections must be made as discreet as possible to minimize the bulkiness of the members.

Architecturally Exposed Systems

The decision to architecturally expose the diagrid will radically alter the decision-making matrix for the project. Although diagrids will naturally require significant custom fabrication, AESS versions are more likely to require custom fabrication for the diagonal diagrid members as well as the nodes. This increases the cost of fabrication.

The AESS diagrid of the Bow Encana Tower in Calgary, AB, Canada, designed by Foster + Partners and Zeidler Partnership, uses a welded joint between the node and the diagonal. As these surfaces are exposed within the atrium, they have been filled and ground after welding so that the transition is invisible once painted. Temporary bolted tabs are used to secure the diagonal prior to welding. The connection is filled and sanded to achieve a seamless appearance.

An architecturally exposed diagrid is more likely to require fully welded connections – both within the node and for the connections between the nodes and the diagonals. This puts added pressure on the fabrication tolerances in order to ensure that the butt-welded surfaces are closely aligned to ensure a seamless transition. Site welding may also require preheating of the material, so scaffolding and weather protection might also be required. The matching tabs that allow the members to be temporarily bolted during erection will need to be removed and the connecting surfaces ground and filled to remove any traces of the temporary steel attachments.

AESS diagrids that use extensive welding will want to maximize the work done in the shop under controlled conditions to minimize site welding. Transportation will require more careful assessment as the sizes shipped from the shop will want to be of maximum size. Staging area on the site will become more important as there needs to be more careful handling of the members to avoid damaging them during the erection process. A larger staging area can permit some preassembly prior to lifting (this holds for concealed systems as well).

Most important if considering an AESS diagrid is to ascertain the impact of the local fire code. These vary substantially from country to country, and what might be permitted in China may not be in North America. The most common method of fire protection for exposed steel is the intumescent coating (with or without an additional fire suppression system). These coatings vary in thickness and finish as a function of the amount of protection required. Generally speaking, thinner steel requires thicker coatings. In all instances, the finish has a texture much like an orange peel and will mask some of the finer welding and finishing details. In high-traffic areas, or where the public may come into contact with the steel, a durable topcoat is suggested to maintain the quality of the finish.

Larger AESS diagrids may use large-diameter hollow members as these can be fire-protected using the method of concrete fill (see page 57).

NODE ADAPTATIONS FOR CONCEALED SYSTEMS

As was evidenced in the node designs for the concealed systems of the Swiss Re Tower and the Hearst Magazine Tower, a combination of shop prefabrication and bolted connections made on site rendered a geometrically complicated structural system more manageable. The node designs used in Swiss Re and Hearst served as precedents for more recent diagrid buildings. Concealed systems have tended toward the use of wide flange members (Universal sections), although there have also been completely custom-fabricated systems used.

The design of the nodes and the diagrid members clearly responds to the loading conditions. Comparing the general appearance of the nodes for the Hearst Magazine Tower, a 46-storey building, with those for the Bow Encana Tower, a 57-storey building, it is obvious that the nodes for the Bow Encana Tower have been designed for higher stresses. The member sizing for the Aldar Headquarters falls below, given its 25-storey height and additional stresses as a result of its curved shape. Comparing these with the geometrically very simple nodes for One Shelley Street, it is clear that the 10-storey height of this structure greatly simplified the node design.

The use of a node is not, of course, limited to pure diagrid structures. Hybrid structures such as the truss tube frame on the north façade of the Bow Encana Tower or the honeycomb structure used in the Al Bahar Towers have adopted nodes to handle similar connection conditions. The benefits of shop-prefabricating such complex connections are quite clear.

The nodes on the north façade of the Bow Encana Tower, which are part of a truss tube frame, use bolted side plates for the member-to-node attachment. This method was developed in the Hearst Magazine Tower. As a truss tube system must incorporate vertical members, the node needs to provide attachment for eight incoming pieces of steel. Therefore, the load transfer is achieved in part through bolting but as well through site-welded connections between the members. Temporary side tabs were used to bolt the elements together in order to permit the quick release of the crane and preparation for welding.

The external diagrid for One Shelley Street, Sydney, designed by Fitzpatrick + Partners, is created using wide flange (Universal) sections. As the climate in Sydney is temperate, the architects chose to place the diagrid outside of the environmental enclosure. Yet as the environment is also extremely corrosive, the members are hot-dip galvanized as well as being clad for a cleaner appearance. As the horizontal ring bracing provided by the floors is hidden behind the cladding, the X-shape of the nodes is clearly expressed. The two diagrid members that are situated below the node are bolted to the node prior to lifting to speed up the erection process.

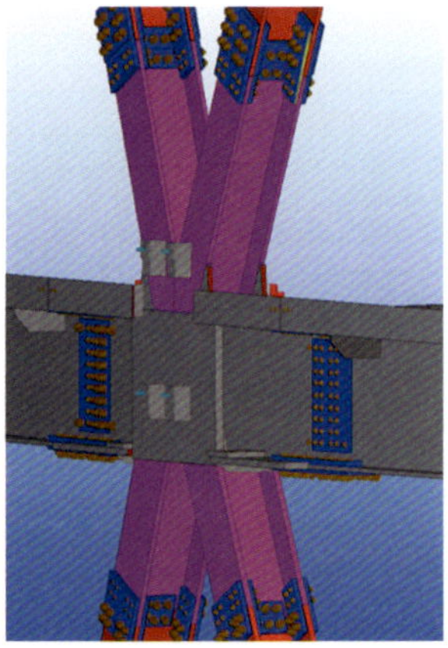

This Tekla Structures detail, prepared by William Hare, shows that the node design for the Aldar Headquarters in Abu Dhabi, UAE, designed by MZ Architects, is similar to the solution for the Hearst Magazine Tower. The node uses wide flange (Universal) members and bolted connections to accept the connecting elements.

This view of the rear of the node for One Shelley Street shows the elements that will provide connection to the floor beams on the interior of the building. All of the on-site connections will be bolted. The members have been sized to suit the capacities of the galvanizing facility (tank size). Galvanizing requires access to all surfaces for dipping and also that the thickness of the steel accounts for the potential deformation from the heat of the bath.

The honeycomb structure for the Al Bahar Towers in Abu Dhabi, UAE, designed by Aedas Architects with Arup, uses steel nodes to complete the connections and load transfers at the points where the vertical columns join with the diagonal members. The nodes allow for clean, bolted on-site connections and cost savings through the prefabrication of the nodes, which also assists in creating tight control on the angle of attachment of the sloped members.

There are concealed structures that do not have a uniform system of diagonals and nodes. This would be the case for buildings with extremely eccentric loading or highly irregular shapes. Although there will be consistency in the approach to engineering the connections, mass prefabrication is not possible as the jigs and setup for each of the elements will be different. However, the advanced BIM and steel detailing software enables the creation of shop drawings for each and every element and connection of the structure. This translates into savings and precision in terms of the alignment and fit for these complex members.

A diagrid structure is used in addition to a column system for the CCTV Building in Beijing, China, designed by OMA. The “butterfly plate” connection type that is used to transfer the loads through the diagonals and into the column system varies throughout the building. The custom-fabricated box sections that are used to create the diagonals are welded to the butterfly connectors. The traces of the temporary plate connection tabs are still visible and will not need to be ground smooth or otherwise remediated, as this side of the structure is concealed.

Even within the same building, concealed diagrid systems may take on very different forms and roles as a function of position and loading. These images of the Addition to the Royal Ontario Museum in Toronto, ON, Canada, designed by Daniel Libeskind, show a more standard connection that sits in the middle of a "face" on one of the crystals (right); and (left) a condition at one of the points of a crystal form that must accommodate multiple convergent members and many different planes.

There are clearly circumstances where the node-and-member strategy is not applicable to the diagrid, for instance in the case of a more chaotic, irregular structure that does not have any modularity, so that most connections are unique. The Addition to the Royal Ontario Museum (ROM) in Toronto would qualify as such a case. Many of the nodal connection points were shop-fabricated as the end condition of the members and transported to the site in one piece. The construction also included straight diagrid elements that were joined to the node end of an element. These were incorporated into the structure as on-site assembly progressed. In many projects larger assemblies are created in the staging area prior to lifting to minimize the number of individual lifts and to fit within the constraints of transportation. More on this strategy will be discussed in *Chapter 8: Constructability*.

NODE ADAPTATIONS FOR ARCHITECTURALLY EXPOSED SYSTEMS

The choice to architecturally expose the steel diagrid system adds a high level of performance to all of the previously mentioned requirements. For extensive information on AESS please refer to Chapters 5 and 6 of *Understanding Steel Design: An Architectural Design Manual* as well as *Architecturally Exposed Structural Steel* (under preparation at the time of writing).

The design of the nodes and members for an exposed system is a critical part of the architectural expression of the space. The AESS diagrids that have been constructed to date have all taken varying approaches to the design of the system, for varying reasons including, but not limited to, scale, height, building use and budget.

When the nodes are expressed in a building that comprises a range of conditions, it will be important to create a consistent language of detailing that can be adapted for the various conditions. These could simply be, for instance, a typical mid-façade condition plus top of building, base of building and corner conditions. Rectangular buildings sometimes present more challenges as these can include different inside and outside corner conditions. Circular or curved plans and sections will also require customization of the nodes, but seldom with the same drastic geometry as an outside corner condition. The designers for the exterior diagrid on the Manukau Institute of Technology in Auckland, New Zealand, developed a comprehensive suite of details to lend coherency to the diagrid design. For example, although the diagrid steel structure is exposed, simple bolted connections are used to attach the diagrid members to the nodes rather than requiring more expensive on-site welding.

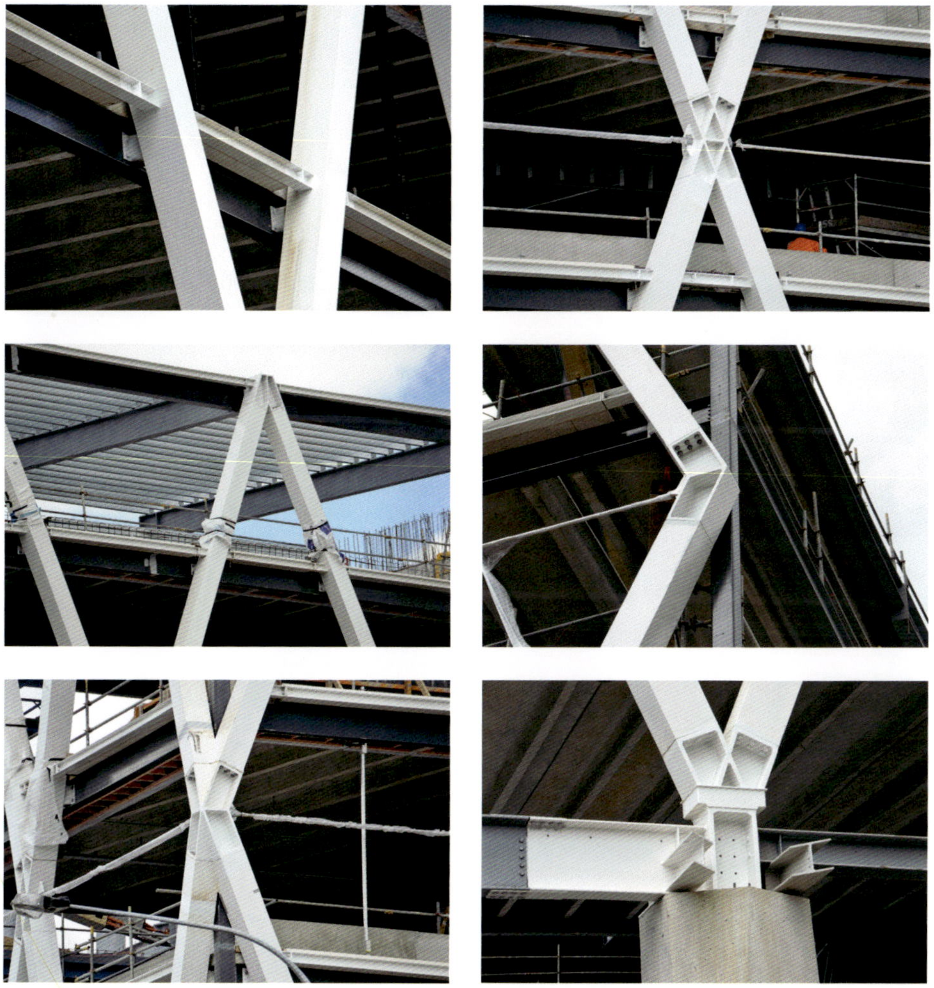

The exterior diagrid system for the Manukau Institute of Technology in Auckland, New Zealand, designed by Warren and Mahoney Architects, uses a coherent language of detailing for the various node conditions throughout the building.
The nodes are all variations of an X-shape. The members are custom-fabricated square box sections. Custom-fabricated platework is often preferred instead of standard HSS members because the rounded corners of the latter do not meet the desire for crisp corners. The horizontal bracing at the floor level uses small wide flange sections to tie in with the visual appearance of the nodes.

The idea of a suite of node variants was also important for The Leadenhall Building, whose asymmetrical design led to differentiated structural conditions on the front, sides and rear of the tower. The simple design approach might look similar to the one taken by the Manukau Institute of Technology, but the design and engineering requirements for the nodes were extremely complex. The trapezoidal form of the building results in a more standard diamond-shaped diagrid system on the front, a diagonal-and-column system on the sides and a rear structure that includes diagonal bracing and columns, with eight members converging at a node. Its 50-storey height, as compared to Manukau's five-storey height, puts significant loading stresses on the nodes. The high loads that must be transferred through the diagonals and nodes required extensive custom-fabricated platework. Although the design of Manukau had to allow for the high seismic loads of New Zealand, tall buildings have much higher wind loads. The wind loading on Leadenhall, combined with the cumulative floor loads and eccentric form, pushed the design strategies for the nodes. Where the low-rise diagrid system of Manukau could use simple bolted connections, Leadenhall uses a tension bolting system as the bolts needed to be adjusted during construction to account for deformations due to changes in loading and column shortening. In some instances the "wide flange" appearance masks the fact that the members have two webs so as to provide additional strength. Even where the diagrid system connects to the chevron-type bracing system at the rear of the building, the connection language remains consistent. This will be discussed in more detail in The Leadenhall Building Project Profile (see page 148).

A coherent language of detailing has been employed for the node designs for The Leadenhall Building. The X-shape of the node is varied to reflect the location and geometric condition. The detailing language is varied subtly for the chevron bracing system at the rear side edges of the building.

Many of the larger AESS diagrids use members and nodes that have been custom-fabricated from plate steel, as the architectural intentions of the project are not satisfied by standard profiles. Higher gravity and lateral loads can also require very large members that might not be satisfied by the load-carrying capacities of standard wide flange or HSS sections. This was the case with The Leadenhall Building, where the sections were built up from welded plate. Welded plate members also have a "sharper" look than hot-rolled members.

The diagrid system on The Leadenhall Building in London, England, by Rogers Stirk Harbour and Partners with Arup, creates a vibrant language of exposed detailing for its diagrid structure as the structure will be exposed on the exterior and interior of the building as it forms part of the double façade envelope system.

Other examples are the nodes and diagrid members for the Bow Encana Tower that were custom-fabricated from plate steel and given a triangular cross section. As the steel diagrid was to be exposed as an AESS structure to the double façade in the atrium, an all-welded system was chosen. This, in turn, required the construction of weather enclosures at each of the nodes to provide the ironworkers with a protected environment to complete their work on this project in Calgary where the weather is extremely cold, snowy and windy in the winter months. Climate and providing proper weather protection and scaffolding access are important considerations for these projects.

The considerations for the fabrication of the nodes and members for the Bow Encana Tower included maximizing shop fabrication and shipping size. The rear face of the node shows a "concealed" level of finish as it will ultimately be hidden by the exterior cladding. The custom steel diagrid was fabricated by Walters Inc.

The Bow under contruction, showing the backside of the diagrid structure that will be hidden. Reducing the finish expectations for this surface saved on project costs.

To achieve some economies on larger systems, it is possible to finish the members according to their uses – that is finish the exposed faces to AESS standards and finish the concealed faces to basic structural standards. While this is likely not to apply to many situations, it is helpful to know that it has been done and is a possibility. The large diagrid façade of the Bow Encana Tower had the interior face of the diagrid members finished to a high level of AESS (AESS4 according to the Canadian system) as they are exposed to the public atrium. The backsides of the members are concealed by a curtain wall system and were more simply finished.

The basic method of fabrication and the connection between the nodes and the connecting members can be relatively consistent even in cases where each node is unique. This is the case in the extremely challenging construction of the node and diagrid system for Capital Gate in Abu Dhabi. Due to the backwards-leaning geometry of the tower, the diagrid needed to be very rigid to restrain the eccentric load. The concrete core of the building did play a big role in this, yet still the diagrid had to use a very small two-storey module.

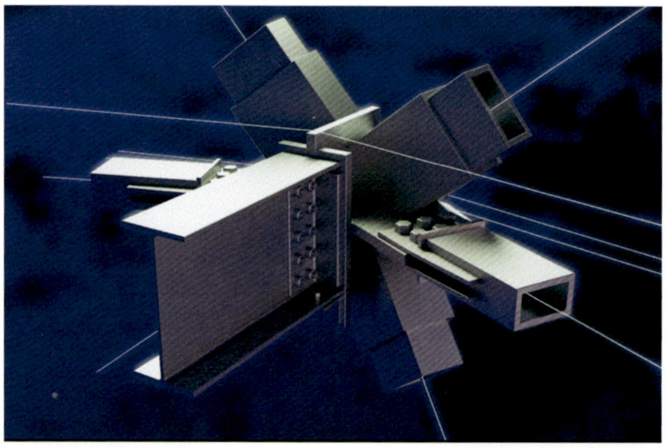

These 3D models of the nodes for Capital Gate in Abu Dhabi, UAE, were created by Jeff Schofield of ADNEC to explore the visual nature of the connections and alignment. The more than 800 node connections on the project were all different. Digital modeling was very important to examine a range of design and construction issues.

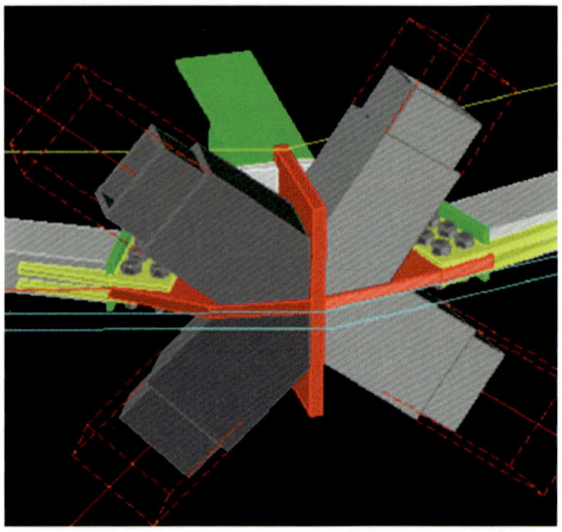

The use of steel fabrication detailing software is essential for complex diagrid systems. This Tekla Structures rendering of a node for Capital Gate shows the way in which the elements are assembled. The horizontal tension braces are attached via a site-bolted connection.
The majority of the nodes make use of a cross-shaped set of plates to create the connection between the incoming tubular members.

Where the diagrid is used in the Capital Gate public areas in viewing distance to reveal the node and its bracing, an alternate design had to be used. In the ground floor lobby (left), a custom cover is used to conceal the bolted connection for the horizontal brace. At the restaurant on the 18th floor, the diagrid member is significantly larger due to the particular loading conditions at this level.

The importance of a consistent language of detailing becomes obvious in the interior atrium of Capital Gate. The use of the cross-shaped plate connector has been maintained, but the members have been changed to more standard tubular sections as the loads are lighter.
The vertical plate has been rounded to acknowledge the round sections. Intumescent fire protection has been used to obtain a higher fire resistance rating and allow exposure of the steel, as these sections would be impractical to fill with concrete.

Fire resistance for AESS diagrids plays a large role in their design and detailing. Where Capital Gate managed full exposure of the diagrid on the interior of the building through the use of an intumescent coating system, the Guangzhou IFC relied primarily on filling its very large tubes with concrete. Yet the two-hour rating required for the combined office and hotel occupancy at the Guangzhou IFC could not be met with the concrete fill alone. An additional layer of fire-protective coating was trowel-applied to the tubes. Additionally, refuge floors were provided, as is now customary for tall towers.

The large nodes for the International Finance Center (IFC) in Guangzhou, China, designed by Wilkinson Eyre Architects with Arup, create a fairly clean X-form.
The tubes that connect the nodes are temporarily attached on site using bolted clip plates. These are removed after on-site welding has taken place. The node itself does not include much of a detail for the attachment of the horizontal bracing ring. As this diagrid is fully braced by each floor along the length of its 27m/88.6ft diagonals (as seen by the image showing the underside of the steel-framed floor system), there was no need to create a larger-sized ring at the node level.

For some more unusual AESS applications, the node and the elements have been combined into one piece to create a more complete system. This path is chosen in many cases to create larger sub-assemblies that can be welded together in the fabrication shop. These also minimize site assembly and speed up erection.

The Canton Tower in Guangzhou, China, designed by Mark Hemel/Barbara Kuit/IBA Architects with Arup, creates sculptural interest in the diagrid structure by refraining from using the nodal elements that typically keep the intersecting steel elements in line, but instead allowing the members to overlap as a way to create depth in the view of the structure. The tubular elements and their connecting pieces were shop-fabricated so that they could be shipped as larger pieces. Note the scale of the workers on the bamboo scaffolding to gain an appreciation of the extreme size of these members. The tubes with the largest diameters were fabricated with all of the attachment elements and connected end-to-end without intermediate connecting pieces. Pieces with smaller diameters were attached on site.

The sculptural ArcelorMittal Orbit Tower, designed by Anish Kapoor and Cecil Balmond with Arup for the London Olympics in 2012, uses a diagrid-like structure where the node and diagonal member are conceived as a modular unit system. Although most of the angles are inconsistent from element to element, the simple bolted-plate on-site connections allowed for sub-assembly prior to lifting and a speedier erection. The welding of the AESS structure was all done in the shop and the elements arrived to the site prepainted – for a more spectacular to view erection.

7
CORE DESIGN

The service core of The Leadenhall Building in London, England, designed by Rogers Stirk Harbour + Partners with Arup, is situated at the rear of the building. The yellow steel-framed structure is relatively structurally independent, providing nothing in the way of traditional core support to resist lateral loads in the building.

This chapter will look at the design of the core of diagrid buildings, predominantly examining the tower typology. The use of a perimeter structural diagrid in a taller building is the most basic application of the diagrid system and the one with the clearest methodology. The cores in taller buildings have historically had a very clear structural purpose, to resist lateral loads. By contrast, the cores in low-rise buildings vary significantly according to the function of the building, its geometry and local material preferences. Where the choice of steel for the diagrid frame is quite clear, the choice of material for the core is less consistent, as it is subject to decisions made to fit a wide range of variables.

Structural material comparison of the 100 tallest buildings in the world per decade from 1969-2010.

This chart, prepared by the Council on Tall Buildings and Urban Habitat (CTBUH), shows the shift in the choice of structural materials over the past six decades. This would indicate a decrease in the construction of all-steel buildings.

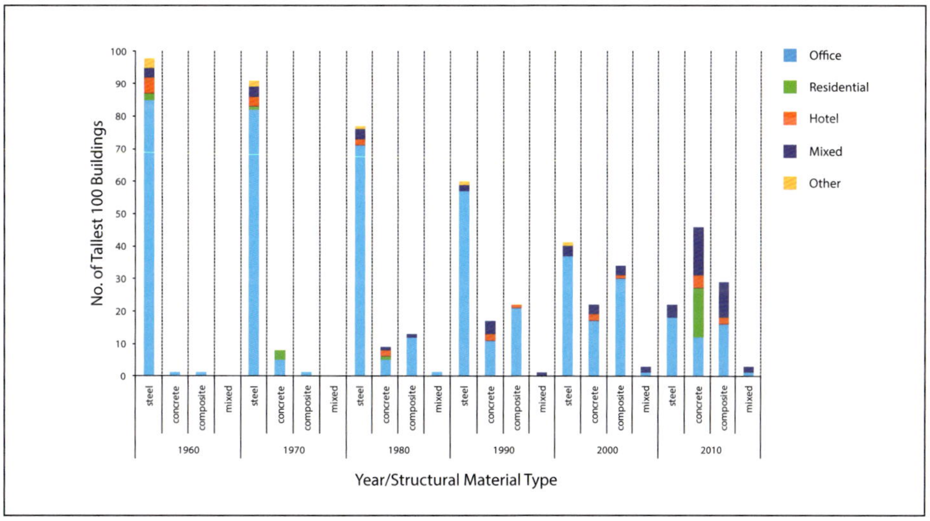

MATERIAL TRENDS IN TALL BUILDING DESIGN

If we look at the history of tall building design in the database established by the Council on Tall Buildings and Urban Habitat (CTBUH), we can see a number of trends that have begun to influence the material choices for the design of tower buildings:

- The construction of tall buildings has shifted away from a North American dominance and to significantly higher numbers in Asia and the Middle East.
- The primary structural material has shifted from steel to composite construction (with concrete coming in second).
- All Supertall buildings are being constructed with composite materials.
- Most buildings use reinforced concrete for the construction of the building core.
- Structural material choices tend to be regional in that complete steel frame buildings tend to be constructed in North America (particularly New York City) and London and not in many other places.
- The price of steel has been volatile due to increasing demand in China, while the price of concrete has been more consistent and reliable.
- Steel tends to require skilled labor, which may not be available.

The first tall buildings were constructed in the United States of America and made predominant use of structural steel for the frame and cores. These early tall buildings were responsible for defining the structural practices for tall building design. It is interesting to note that structural progress in the design of steel skyscrapers was reflected in a confidence in high-strength steel, structural engineering and construction that led to the elimination of redundant systems.

Skyscraper construction did not change substantially until the construction of the 100-storey John Hancock Center in Chicago, designed by SOM, in 1965, that introduced the braced tubular cantilever (also called diagonalized core system). In order to permit larger expanses of glass and less frequent vertical columns, large diagonal members were overlaid to the rectilinear grids of strip windows, column covers and spandrel panels to brace the entire length of the structure.

The diagonalized core typology began to engage the perimeter structure in sharing the lateral loading. This was an important advance as buildings were growing taller and needed to resist higher lateral loads. Lateral loads always include wind loading and may also include varying seismic loads as a function of the location and assessed risk. Historically tall buildings, because they act as vertical cantilevers, have grown to rely on the stiffness of the core to resist lateral loads. While braced steel cores could "hide" the diagonal reinforcing in the opaque elements required to house elevators and stairs, reinforced concrete was naturally resistant and stiff and was easily used to structure the typical programmatic requirements of the core.

A vertical column system was employed to carry pure gravity loads throughout the usable floor area of the building. Over the years the preference for column-free interior spaces has resulted in towers comprised of cores and perimeter support systems. The John Hancock and Willis towers in Chicago and the destroyed World Trade Center towers in New York City are, respectively were, the tallest all-steel towers constructed to date. This includes using steel to frame the core.

THE IMPACT OF 9/11 ON CORE DESIGN

The tendency toward the use of a concrete core has also increased as a result of the terrorist acts that destroyed the World Trade Center towers in New York City. The major question arising out of 9/11 focused on the possibility of designing a skyscraper strong enough to resist a similar attack. Buildings to this point had been predominantly designed to resist natural forces such as high winds and seismic events, but never "acts of war" or manmade events. Although a report prepared by the National Institute of Standards and Technologies (NIST) makes a multitude of safety recommendations for future skyscraper design, two stand out as having the most impact on future progress in American skyscraper design. Firstly, that public officials and building owners will need to determine appropriate performance requirements for buildings that are at higher risk due to their iconic status, critical function or design and secondly, to adopt and use the "structural frame" approach where structural members connected to the columns carry the high fire resistance rating of the columns.[1] In response to the suggestions, The Port Authority of New York has stated regarding the construction of the new One World Trade Center tower:

The diagonal geometry of the new One World Trade Center, designed by SOM, relies on steel framing and a steel bracing system.

"New safety features will include 3 feet (91 cm) thick reinforced concrete walls for all stairwells, elevator shafts, risers, and sprinkler systems... This structure is designed around a strong, redundant steel moment frame consisting of beams and columns connected by a combination of welding and bolting. Paired with a concrete-core shear wall, the moment frame lends substantial rigidity and redundancy to the overall building structure while providing column-free interior spans for maximum flexibility."

The move toward the construction of concrete cores as a result of 9/11 was more significant in the USA than in other places. Up until this point many American tall buildings continued to be constructed in steel. Other parts of the world, Asia and the Middle East in particular, had already begun to boom and had assumed composite construction (with steel and concrete) or concrete construction as their chosen method. They were not necessarily turning away from steel, but rather their emerging developments used other construction methods from the beginning.

The impact of the destruction of the World Trade Center towers on the construction of tall buildings in New York City and beyond has been significant. This has meant a shift from a steel-framed core to a core reinforced with concrete in most cases. The core of the new One World Trade Center tower was constructed in steel and encased in close to 3ft/91cm of concrete for added protection.

The current focus on building increasingly taller in the tall building community, the impact of the realities of 9/11 and the recommendations of various studies and reports have tended to support the choice of concrete reinforcement for building cores, often without a steel frame. Interestingly, the first two diagrid high-rise buildings, Swiss Re Tower (2004) and the Hearst Magazine Tower (2006) both have steel-framed building cores. The design decisions for Swiss Re would have been substantially complete at the time of 9/11 and therefore not impacted. The height of the Hearst Magazine Tower and its location in the city would likely not have been perceived as putting the building in a position of risk. It was the first New York tower to be completed post-9/11.

THE FUNCTION OF THE CORE IN A DIAGRID BUILDING

One of the primary benefits of using a diagrid system for the perimeter framing of the building is its ability to resist the majority of the gravity and lateral loads. The lateral loads can include wind and seismic loads. The magnitude of these loads is determined by local building codes, climate data and specific wind testing, as addressed in more detail in *Chapter 4: Technical Requirements*. This diminishes the dependency on and function of the core in creating the required overall stiffness of the building to resist lateral loads.

While tall buildings that employ other structural methods such as a diagonalized core system, an outrigger system, a bundled tube system or a megaframe will depend on the core for lateral stability, a diagrid tower does not necessarily need to depend on the core for this stability. The complete shift of the lateral load resistance from the core to the perimeter diagrid was established in the engineering design of Swiss Re. This is noted by Dominic Munro of Arup in an article written for the Norwegian journal *Stålbyggnad,* describing in detail the innovations in the design of Swiss Re. These considerations lead to the realization that a genuine diagrid building does not need a reinforced concrete core, because such a core will invariably assume the lateral load resistance.[2]

Whether or not the engineering of a perimeter tube diagrid tower chooses to apportion all of the lateral loads on the diagrid, or share the loads with a reinforced core in either steel or concrete, will be a project-specific decision that responds to loads, geometry, height, locally available materials, regional preferences and budget. This means that the construction and role of the core in a diagrid tower can involve different choices and its materiality need not be a given.

One of the technical issues associated with the dependence or not on a core for stability is seismic design. Current seismic codes have been established based on the actions of the core in stabilizing the structure during a seismic event. A perimeter diagrid structure that is assuming all of the lateral loading is functioning as a bearing wall type system and is therefore not addressed in the current seismic codes and practices. This issue is discussed further in *Chapter 4: Technical Requirements*.

STEEL-FRAMED CORES

To the date of writing there have been six notable diagrid towers designed with steel-framed cores: IBM Building, Swiss Re, Hearst Magazine Tower, CCTV Building, Bow Encana Tower and The Leadenhall Building. Each building represents a significantly different case and with varying reasons for the selection of a steel-framed core. The IBM Building, for example, was constructed in Pittsburgh, in a part of the USA renowned for steel production. The year was 1963, terrorism was not a factor and the project followed standard Northeastern US construction practices.

When deciding upon a steel-framed core there are several issues that must be addressed:

- → Lateral loads: If using an unbraced steel core, such as the one used in Swiss Re, the perimeter diagrid must be designed to assume all of the lateral loads.
- → Fire protection: Fire engineering must provide for protection of the core and the safe egress of occupants. While this is an obvious safety require-

ment for all buildings, there may be jurisdictions that will require concrete and will not permit the use of fire-rated gypsum board, for example. Some buildings in particular might need to have disaster mitigation strategies.

- Constructability and erection sequencing: The erection sequencing for an all-steel building is different from that for one with a concrete core. An all-steel building will proceed typically floor by floor, including the core. The crane locations may need to be adjusted if the contractor normally uses a climbing crane that is located, for reasons of stability, in the concrete elevator core.
- Local practices or preferences: Diagrid buildings are unusual and local practices or preferences may overrule. This would include availability and cost of materials.
- Eccentric loading: A braced core may be required if the loading is highly eccentric, as the perimeter diagrid may be structurally insufficient to resist extreme loads. This was the case for Capital Gate (see page 105); however, the case of CCTV would demonstrate that the all-steel solution can be viable even with extreme geometries.
- Building height: Perimeter diagrids that do not use a braced core tend to be limited in height. This is a function of a combination of building height, proportions and lateral loads. The specific limit has yet to be determined.

Centered Steel Core

Swiss Re took full advantage of its perimeter diagrid and open core in its planning. The desire to have a very open plan was a driving factor for its decisions regarding the choice for an all-steel solution. Steel is also a very common structural material in London, so there is sufficient expertise to design and construct innovative solutions in steel. Last not least, Foster + Partners have a long history of innovative steel design, starting in the High Tech Movement.

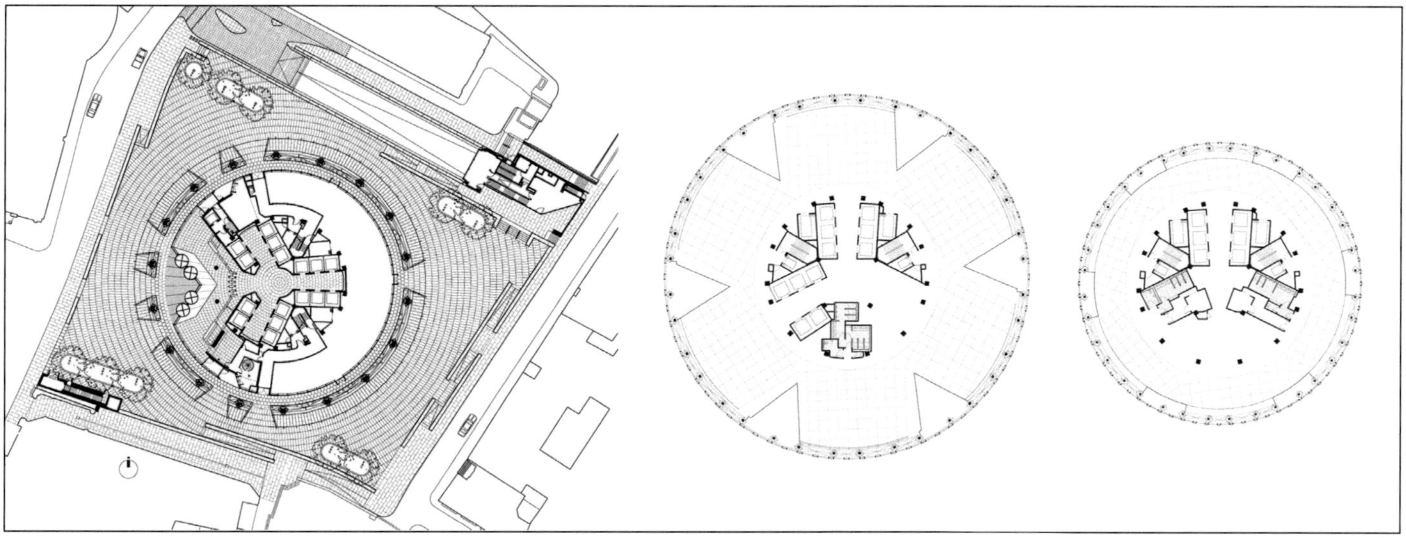

The ground floor plan (left) of the Swiss Re Tower (30 St. Mary Axe) in London, England, designed by Foster + Partners and engineered by Arup, shows the core at its most dense iteration. As one progresses up the building (middle) and toward the top (right), the number of lifts decreases and the floor plan is permitted to open up. The steel framing that contains the core is less intrusive to the planning and feel of the spaces.

Offset Steel Core

The Hearst Magazine Tower, being situated in New York City where all-steel buildings were the norm at the time, and coming from the offices of Foster + Partners, who were used to designing in steel and had just completed Swiss Re using a diagrid, has a steel core as well. Significantly, in the case of Hearst, the core functions were pushed toward the rear of the building as the other three sides had street frontages that permitted better views and glazing, and the rear side did not. That meant that the core would be located eccentrically and not well positioned to assume the lateral loading. The required eccentricity of the core supported the decision to use a perimeter diagrid, as the diagrid could assume the lateral loads. In this instance, the steel core was encased in concrete at the lower levels of the building (within the renovated and reused portion of the historic Hearst building) and all other floors above were constructed with steel framing and no concrete encasement. On the upper levels the steel core was braced with steel as is more traditional for New York City construction, to assist in its participation in the stiffening of the overall structure.

The diagrid tower portion of the Hearst Magazine Tower was set back from the position of the historic façade. The megacolumns that support the diagrid tower are evident on the ground floor plan, as is the concrete core. The typical upper level plan shows the steel-framed core, which allowed for a more flexible plan.

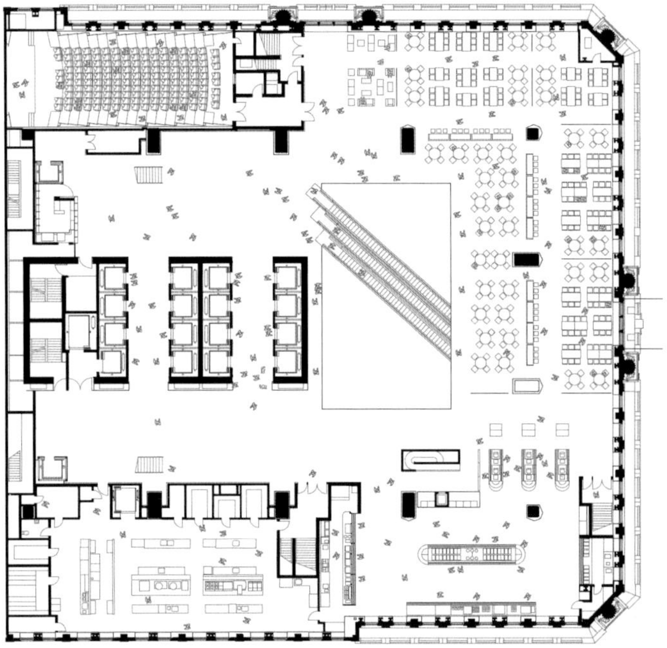

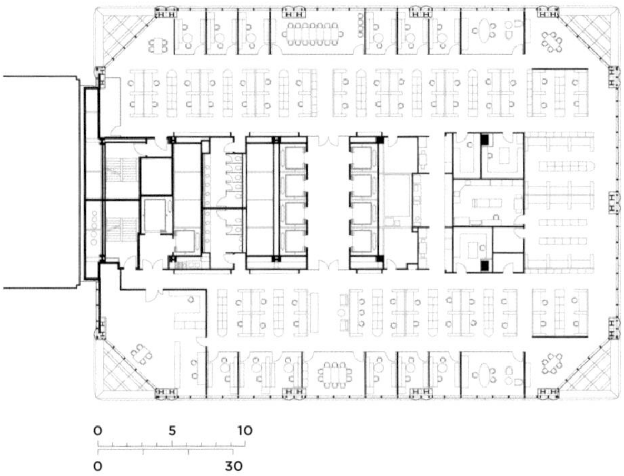

Steel Core Outside of the Building

Where the steel-framed core for Swiss Re was located in a traditional center location, and Hearst slightly offset, the core for The Leadenhall Building is actually positioned behind the main diagrid tower structure. This is likely the tallest example of the building core being so completely disengaged from the structure and clearly relieved of its "normal job" of being the primary source for the resistance of lateral loads. In this instance, the ability of the diagrid structure to be capable of resisting the lateral loads has been fully exploited as well as exposed. The actual function of the Leadenhall diagrid will be more fully explained in its Project Profile on page 148.

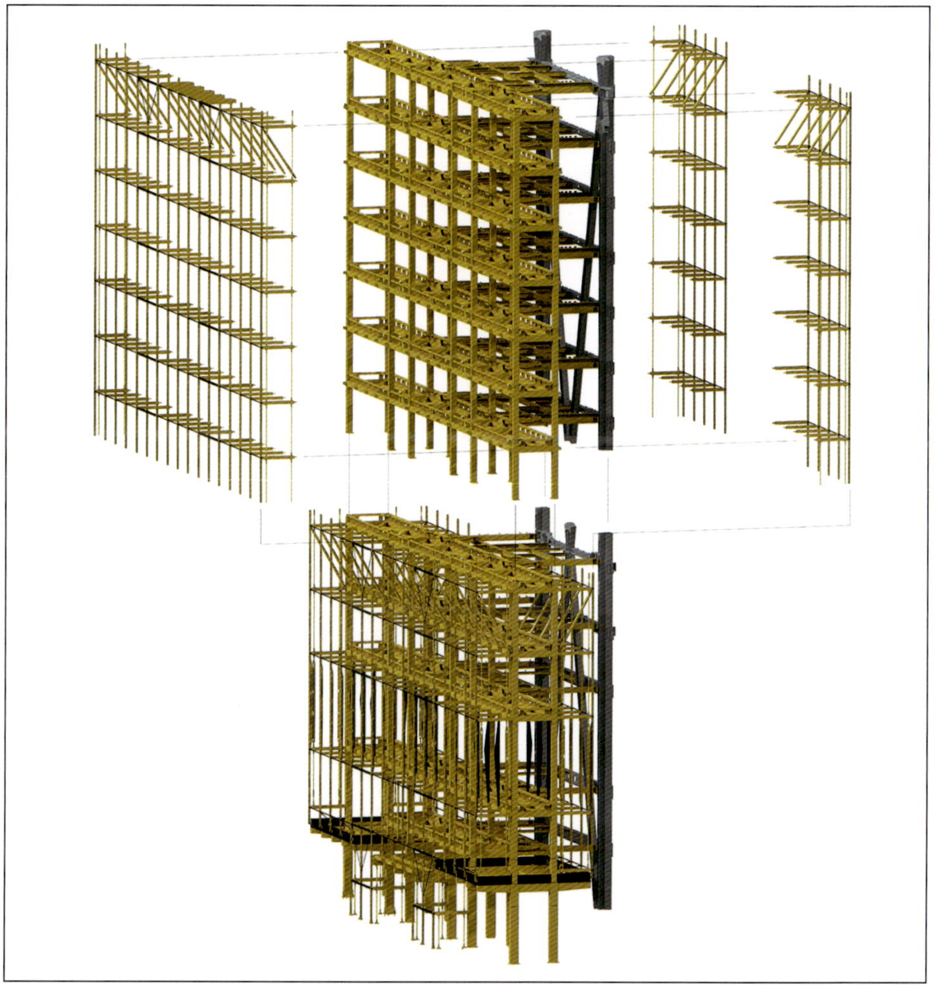

The yellow-painted AESS steel core for The Leadenhall Building is located at the back of the building. Its spaces are linked to the main office floors by a steel-framed lobby. The core does not assist the primary diagrid structure with lateral loading. The accuracy of the digital model and the current structural and fabrication software and methods results in a very high level of consistency from the design through fabrication stages.

The steel-framed core of The Leadenhall Building continues Rogers' AESS theme in terms of using the service core as an expressed feature of the building.

The Leadenhall Building is also clearly a project where the architectural and engineering team have a very high level of expertise and comfort in exploring more daring structural arrangements in addition to a long history of detailing in AESS. This is also a project team that has fully engaged its steel contractors and detailers in terms of the sheer amount of Architecturally Exposed Structural Steel materials and systems. This requires skilled coordination and extreme care in handling.

Steel Cores for Hybrid Diagrid Buildings

Some applications of diagrids use the system in atypical ways. This would be the case for the Bow Encana Tower in Calgary and the CCTV Building in Beijing. Both buildings use extremely complex structural systems, including a diagrid in addition to other steel framing methods. The buildings have little more than this in common, as their geometries and reasons for the selection of their steel systems were very different.

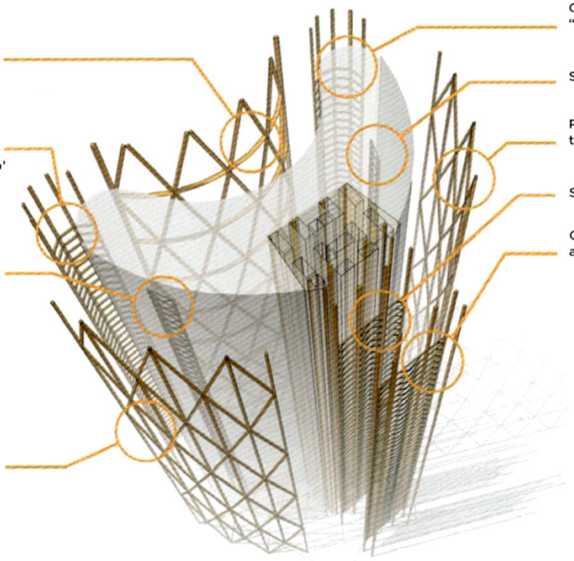

This diagram of the Bow Encana Tower in Calgary, AB, Canada, designed by Foster + Partners with Zeidler Partnership, illustrates the number of structural steel systems that are used to stabilize the building. The core is framed in steel and is not relied on for the resistance of lateral loading.

The Bow Encana Tower uses a number of steel framing solutions. A pure diagrid is used on the south façade to form the primary atrium wall.

Unlike some of the more pure diagrid structures, the Bow Encana Tower has a fairly large floor plate. It uses a diagrid only on its south façade, as a means to enclose a large multi-storey atrium space. Columns and beams structure the majority of the interior spaces. The steel-framed core is not relied on for lateral stability.

This diagram of the CCTV Building in Beijing, China, designed by OMA with engineering by Arup, shows the distribution of the public spaces and circulation in the building. The eccentric geometry of the building would have made the use of the cores to balance the loads extremely difficult. The diagrid overlay was a highly innovative means of solving the loading problem.

The diagrid framing is laid over a system of beams and sloped columns. As the floor plate areas are very large, a system of interior columns is also used.

The massing of the CCTV Tower did not allow for the creation of a typical core system.

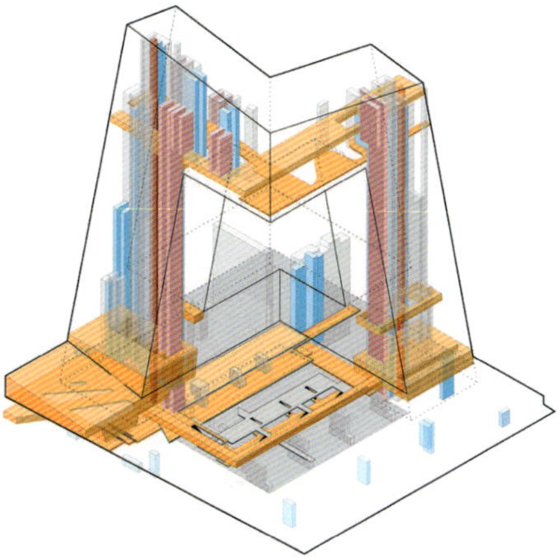

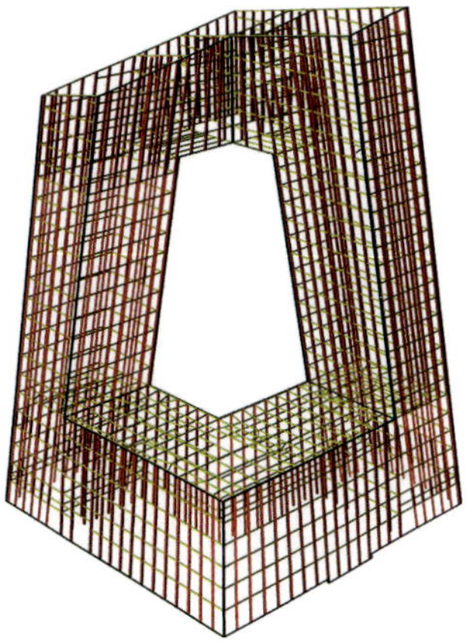

The structural system of the CCTV Building in Beijing was the result of an extraordinary effort by the OMA and Arup team: a building with this sort of eccentric geometry and loading had never before been attempted. The diagrid was used in part to permit the construction of the large multi-storey cantilevered sections of the tower without need of temporary support. As can be seen from the overall structural diagram, the diagrid was applied over a more regular framing system comprised of beams and sloped columns. The density and pattern of the diagonal system varies according to the loading. The core is framed in steel and not used to resist lateral loading, in part because the specific tasks created by the form of this building exceed the ability of a core. The location of the service core as it would work for the planning of the building would not have put it in a position to help with the lateral loads created by the large cantilevers.

REINFORCED CONCRETE CORES

Reinforced concrete cores are more the standard for tall buildings in general, although, as has been mentioned, a pure diagrid building does not necessarily require the assistance of the core to resist lateral loads. As will be demonstrated with some examples where concrete cores have been used, tall towers can require the assistance of the core. Concrete cores are often chosen for other, practical reasons that may lie outside of design-driven motivations. When deciding upon a reinforced concrete core there are several issues that must be addressed:

- Lateral loads: By its nature, a reinforced concrete core will assume the resistance of lateral loads. This can be used to assist the perimeter steel diagrid.
- Fire protection: Many jurisdictions and codes may require that the service core of the building be constructed from reinforced concrete for matters of fire and emergency safety.
- Constructability and erection sequencing: The erection sequence for a building with a concrete core will see the casting of the core proceed ahead of the steel framing. The core and elevator shafts can therefore allow for the use of a climbing crane rather than having to use separate tower cranes. The core will have steel insets to provide for the attachment of the steel frame.
- Local practices or preferences: Concrete cores may be the preferred local practice. Concrete may be chosen due to material availability and cost.
- Eccentric loading: A braced core solution may be required if the loading is highly eccentric, as the perimeter diagrid may be structurally insufficient to resist extremely eccentric loads. The bracing can take different forms as a function of the building design and loading. This can be the case for extremely thin buildings, such as the Aldar Headquarters (see page 104) where the shape of the plan would make it difficult for the diagrid to resist all of the wind loads as they strike the wide face of the building; or in the case of Capital Gate where the loads are eccentric and the tower has relatively small floor plates (see page 105).
- Differential movement: When combining steel and concrete, the tendency of concrete to creep or shrink over time must be balanced with the tendency of steel to be subject to high rates of thermal expansion.
- Building height: Reinforced concrete cores tend to be used in very tall buildings where the lateral loads are high and the core is needed as an additional lateral load-resisting system; i.e. the perimeter diagrid is not able to provide all of the resistance.

The Royal Ontario Museum by Daniel Libeskind incorporated a concrete core into the building. Although the initial diagrid elements used the core for stability during contruction, much of the completed complex structure did not rely on the core to assist with overall stability.

The attaching plates for the steel beams that will clear span from the concrete core to the perimeter framing must be carefully integrated into the casting of the concrete core as it proceeds. Here you can see that the round shape of the Al Bahar Tower is created through faceting rather than bending the steel. Steel fabrication and erection by William Hare Limited.

The Al Bahar Towers in Abu Dhabi, UAE, designed by Aedas Architects and engineered by Arup, employ a concrete core design in conjunction with a steel honeycomb frame. Concrete is a very dominant building material in the UAE, with steel seeing less frequent use. The dynamic mashrabiya shading system creates some eccentric loading on the structure.

Centered Concrete Core

The centered core is the most typical plan arrangement. It results in little or no eccentric loading and the core and perimeter framing can evenly share the lateral loads of the building. There may be some instances where wind studies indicate that the loads are not evenly applied. The Al Bahar Towers in Abu Dhabi, designed by Aedas Architects with engineering by Arup, although using a honeycomb adaptation of a diagrid, use the concrete core in a manner quite representative of this tower type, particularly in terms of the relatively small floor plate and the desire for column-free space between the core and the perimeter support system. In the case of Al Bahar there would be some eccentricity in loading due to the use of the movable mashrabiya, which opens and closes according to the sun exposure. This puts additional weight on the east, west and south façades, the north façade having its glazing left clear. Reinforced concrete and composite constructions are the dominant building materials in the Middle East and this would also be a factor in the decisions surrounding the construction of these towers.

Concrete Core for a Narrow Plan

An extremely narrow floor plate will create eccentric loading, particularly in the case of wind loads that strike the larger face. This may create an overturning moment that may not be easily solved by simply using the perimeter diagrid of the building to resist the lateral loads. The Aldar Headquarters in Abu Dhabi is the only circular or disk-like building in the world. At 110m/361ft it would be classed as a tall building, although in its proportions is does not present itself as particularly tall. It has a steel diagrid to frame each of its large round faces. The building has an extremely narrow floor plate and the use of the paired concrete cores is able to assist in the resistance of lateral loads. Aldar does not use a diagrid to connect the two faces.

The design of the Aldar Headquarters in Abu Dhabi, UAE, designed by MZ Architects with engineering by Arup, is highly specialized, creating an extremely small floor plate, especially if looking at the core-to-floor area ratio. The narrow plan results in high levels of daylighting in the office areas. The core has been divided into two components to better fit the plan and resist loading.

Concrete Core for Highly Eccentric Loading

Capital Gate in Abu Dhabi, designed by RMJM and self-engineered, is the furthest backward-leaning tower in the world - at 18° it far exceeds the eventual lean of the Leaning Tower of Pisa. The previous maximum lean accomplished in the modern era was for Puerta de Europa Towers in Madrid, designed by Philip Johnson and John Burgee and engineered by Leslie E. Robertson. These towers are diagonally braced to counterbalance their 15° lean; they also rely on concrete cores and a system of post-tensioning. They were completed as the first leaning office towers in 1996 and stand at 114m/374ft.

The Puerta de Europa Towers in Madrid, Spain, designed by Philip Johnson and John Burgee, use their mirrored 15° lean to frame an important entrance to the city. The detailing on the façade highlights the precise offset as based on the accentuated vertical mullion.

The structural concrete cores and the steel frames were post-tensioned to compensate for the cantilever of the buildings. A concrete counterweight, located on the opposite side of inclination underground and connected to the top by cable, provides the necessary resistance to counteract the forces trying to overturn the towers.

At 165m/540ft Capital Gate would be classed as a tall building. Initiated by the Abu Dhabi National Exhibitions Company (ADNEC), who wanted a signature piece, the building is unusual in that it has a perimeter diagrid as well as an interior diagrid that creates an upper-level atrium. The diagrid and the reinforced concrete core were carefully engineered to work in concert to resist the loads.

According to an article written for the *CTBUH Journal* by Jeff Schofield, the concrete core occupies the only continuously available space in the tower profile.[3] It has a high amount of reinforcement especially at the points where the steel frames into the core to ensure stiffness and load-sharing. During construction the core was poured "pre-cambered" by 350mm/1.12ft so that as the eccentric loading of the floors was added, the core straightened out. This was essential to provide proper alignment for the elevators. The core was also post-tensioned with vertical cables installed only on one side to counteract the lean on the other side. These tension cables were installed in vertical segments that overlap each other every seven floors. Strung together, they span the entire height of the core to maintain it in a perfectly vertical position.

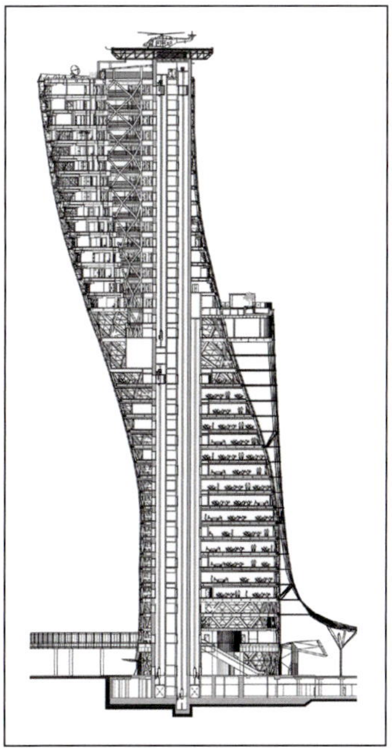

The cross section of Capital Gate in Abu Dhabi, UAE, designed by RMJM, shows the alignment of the concrete core. The core rises vertically through the tower, with the proportion of floor loads shifting dramatically from one side to the other of the tower. The upper floors, which house a high-end hotel function, have at their center a diagrid-framed atrium. This structurally separates the perimeter diagrid from the core. The interior diagrid that creates the upper-level atrium is structurally linked to the core by eight unique structural pin-jointed members.

During the core construction at Capital Gate, the pouring of the core preceded the installation of the steel floor beams and diagrid. The core was used to counterbalance the significant cantilever of the floor areas and masses.

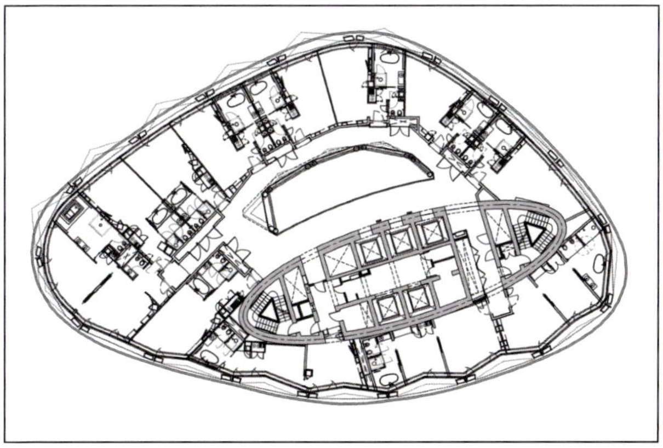

The floor plan of Capital Gate on the atrium level responds to the eccentricity of the building. The core maintains very clean lines and is a tightly organized element.

Concrete Cores for Supertall Buildings

If tall buildings have a higher lateral load to contend with, Supertall buildings must achieve even greater stiffness to resist wind and seismic loads. The Council on Tall Buildings and Urban Habitat (CTBUH) defines a "Supertall" as a building over 300m/984ft in height, and a "Megatall" as a building over 600m/1,969ft in height. At the time of writing no diagrid structures have been built or proposed that would fit the Megatall category. Two visionary projects, Zhonggou Zun by TFP Architects (528m/1,732ft) and the Lotte Super Tower by SOM (555m/1,819ft), fall short of the criteria. The Canton Tower designed by Mark Hemel/Barbara Kuit/IBA Architects with Arup is 600m/1,969ft tall, but is classed as a telecommunications tower due to its proportion of usable area to core, so does not qualify as a Megatall building. It does use a reinforced concrete core to assist with lateral load resistance. The Canton Tower's steel diagrid is more complicated than many in terms of the highly variant conditions of connection between the tubes. This is as a result of the choice to spiral and overlap the tubes rather than having a symmetrical system with in-line nodal connections.

The construction of the reinforced concrete core for the Canton Tower proceeds ahead of the erection of the complex tubular diagrid.

The exterior steel diagrid of the Canton Tower is tied to the reinforced concrete core via deep steel beams. These are also used to frame the support system for the floor areas. One of the challenges in terms of detailing, fabrication and erection was the spiraling configuration of the diagrid system.

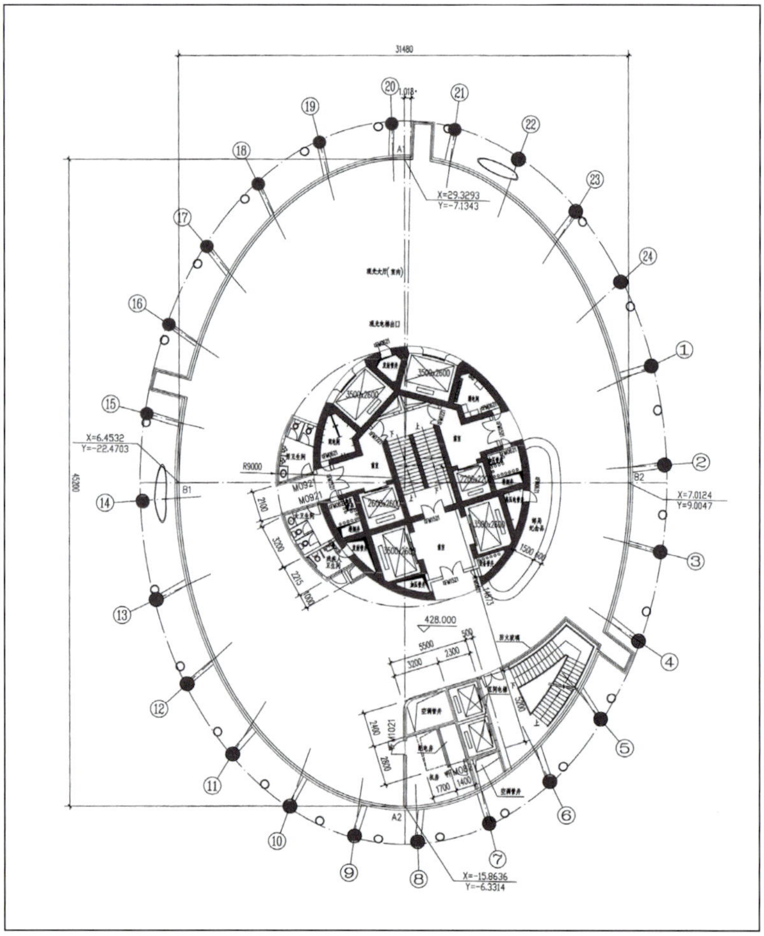

The plan of the observation level floor at 428m/1,404ft of the Canton Tower in Guangzhou, China, designed by Mark Hemel/Barbara Kuit/IBA Architects with Arup, shows the relationship between the reinforced concrete core and the usable floor area. The diagrid frame (concrete-filled tubes) sits outside of any environmental enclosures, subjecting it to thermal movement.

A building of the height of the Guangzhou IFC uses a combination of a perimeter diagrid and a reinforced concrete core to resist lateral loads. The selection of reinforced concrete for the core also works toward the fire engineering strategies of the building.

The Guangzhou International Finance Center (IFC) in Guangzhou is the tallest diagrid tower constructed to date. At 439m/1,439ft it qualifies as a Supertall building. The tower had to undergo rigorous design for wind loading as well as respond to seismic issues. The plan was designed as a curved trochoidal shape and uses a perimeter diagrid of concrete-filled steel tubes in conjunction with a reinforced concrete core to satisfy the lateral and gravity loads. Chris Wilkinson stated[4] that this arrangement to accommodate wind and seismic loads worked so well that the large 30-storey interior atrium at the top of the building could be built without requiring any additional bracing or damping devices. This solution for achieving the desired column-free spaces was more elegant than the outrigger systems normally used for buildings of this height.

Towers of this height cannot use standalone tower cranes to assist with construction. The crane system is normally integrated into the construction of the concrete core as it proceeds ahead of the steel lifts. The central core of the IFC that services the office spaces on the lower 70 floors of the building was constructed using a "climbform" system.

Tall buildings need to have greater moment resistance and stiffness toward the lower half of the building and less toward the top, where the shear loads tend to be higher. This means that the central concrete core may even be discontinued toward the top of a tall building. In the case of the IFC, the primary reinforced concrete core terminates at the hotel lobby level where the center of the building transforms into an atrium void. The elevators and stairs are split and shifted away from the central area.

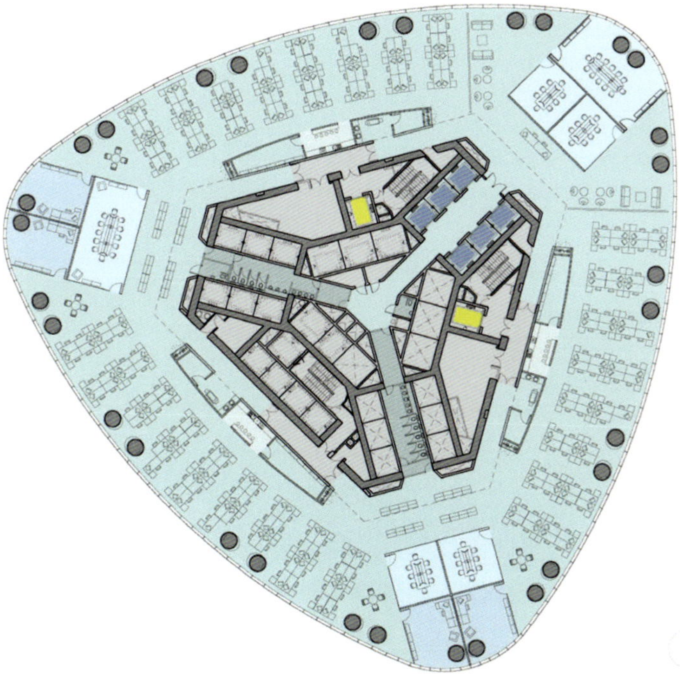

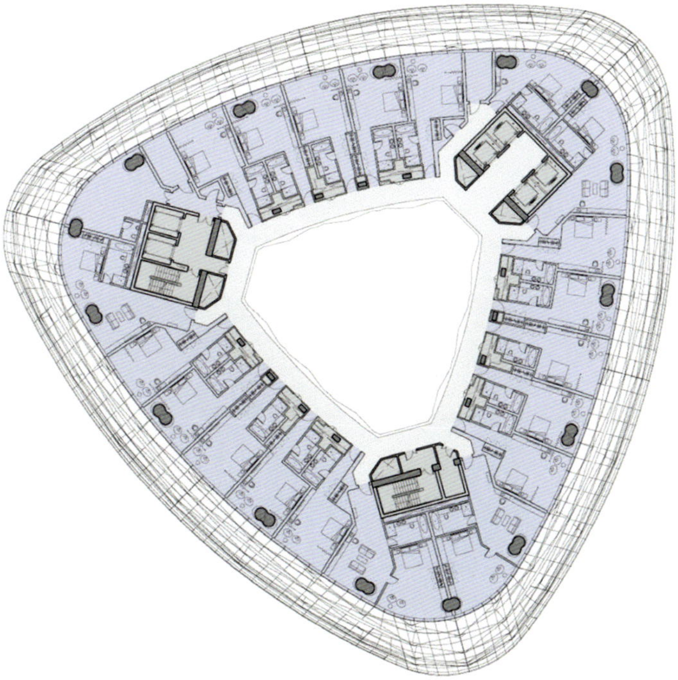

The trochoidal plan of the Guangzhou IFC at a typical office level (left) shows the reinforced concrete core at the center and the column-free office space between the core and the perimeter diagrid. At the upper hotel levels (right), the stair and elevator functions have been separated and occupy the corners of the plan, creating a central atrium space. The atrium is structured by an internal diagrid system that is lighter in scale than the perimeter diagrid system.

Cores in tall buildings undergo very strict review for space planning, as any inefficiency will multiply quickly and result in lost rental revenues. In the IFC an average net-to-gross floor area of 70% has been achieved for the office floors, with a score of 68.9% overall. The triangular plan of the building allows for subletting of up to three separate tenants on each floor; washroom facilities could be tucked into the corners quite neatly. The IFC has 71 elevator lifts, of which 52 serve the office floors, 15 serve the hotel and four serve the below-grade car park.

At the upper level of the Guangzhou IFC, the central core splits and the elevators and staircases move to the sides to make room for a large atrium.

As a rule, the realities of extreme height will always require the use of a structural core, normally in concrete, that acts in conjunction with the lateral strength of the perimeter diagrid to resist loads. This is the rationale behind the advanced structural design of the Lotte Super Tower for Seoul, designed by SOM. An interior reinforced concrete core wall system complements the structural steel diagrid from the foundations through the 112th floor. The core's reinforced concrete walls are strategically configured to optimize material, eliminate transfer zones and provide easy transiting of the tapering floor plates. In addition, the core wall system would provide excellent damping in the overall system to limit wind-induced motions and accelerations.

A rendering of the observation level of the proposed Lotte Super Tower in Seoul, South Korea, indicating the design freedom to be enjoyed when the concrete core can terminate at a lower level.

The design of the Lotte Super Tower also highlights the important role that the floors play in such a composite approach in terms of their connecting role between the exterior and interior systems. According to SOM, long-span floor framing is used to increase the gravity load on the exterior diagrid framing. The diagonals of the exterior diagrid framing are rigidly connected to the major horizontal spandrel beams that ring the perimeter at each floor level and participate in the lateral system by transferring the lateral loads between the diagrid and the floor slabs. The floor framing system for a typical level would consist of a composite metal deck slab with normal-weight concrete topping. The slab would be supported by wide flange structural steel floor framing, which would span between the concrete core wall and the perimeter diagrid. A perimeter structural steel spandrel beam would be connected to the diagrid diagonals and transfer gravity loads from the floor to the diagrid. Within the core, one-way reinforced concrete slabs would span between reinforced concrete beams and girders.

Although this Lotte Super Tower design is not to be constructed, the detailed work on this project is being used to inform other towers designed by SOM that are in preparation for construction.

The strength of the Lotte Super Tower for Seoul, South Korea, designed by SOM, rests in careful design that uses the combined benefits of the diagrids and the core to achieve a 27% reduction in the amount of structural steel on the project, compared to a conventional core and perimeter moment frame solution. The proportion of materials is reduced toward the top of the tower.

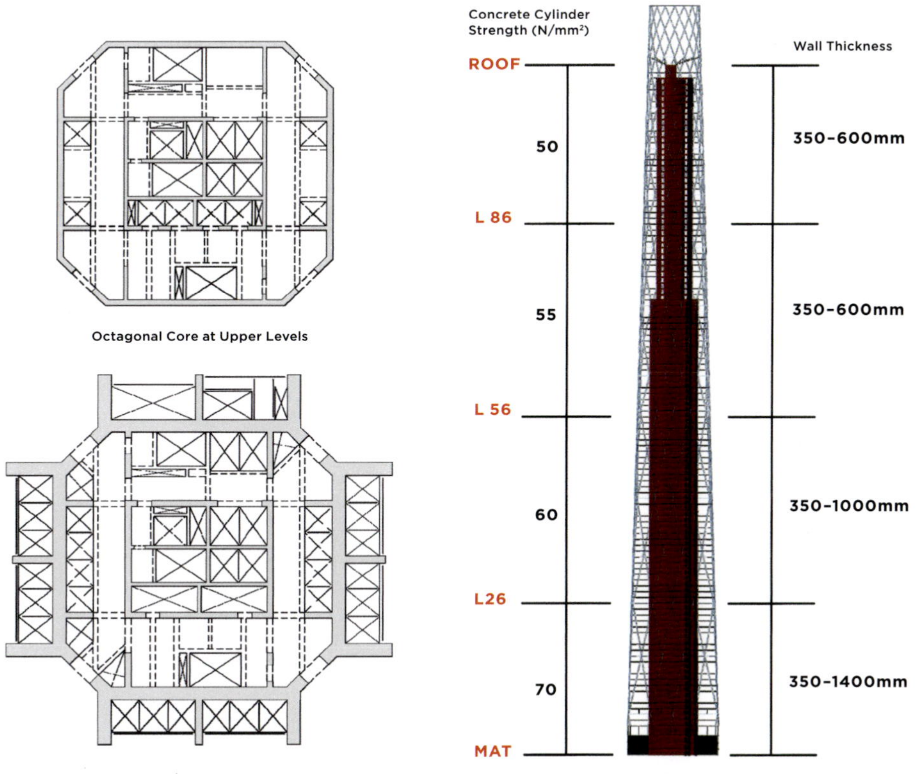

Octagonal Core at Upper Levels

Octagonal Core with Fin Walls at Lower Levels

Concrete Strength and Core Wall Thickness

What becomes clear from the design for both the Ghangzhou IFC and the Lotte Super Tower is that the reinforced concrete core is not required to assist the perimeter diagrid in providing lateral support toward the top of the tower. From a design perspective, this permits "something special" to happen at the top as the center of the building need not be devoted to the function of the core.

8 CONSTRUCTABILITY

SAFETY ISSUES

ARCHITECTURALLY EXPOSED VERSUS CONCEALED STEEL

ECONOMY THROUGH PREFABRICATION AND REPETITION

IMPACT OF NODE AND MODULE CHOICES ON ERECTION

TRANSPORTATION ISSUES

SITE ISSUES

MAINTAINING STABILITY DURING ERECTION

Capital Gate in Abu Dhabi, UAE, designed by RMJM, took the questions surrounding constructability to the highest level of challenge to date for a diagrid building. The extreme complexity and variability of the geometries created many anomalies in erection procedures.

As with any deviation from standard framing techniques, constructability is an important issue in diagrid structures. While more regularized tower projects will be more in line with standard structural steel construction, highly eccentric diagrid buildings will experience significant constructability issues. These must be clearly discussed by the project team in the design phase in order to properly anticipate the impact of the accommodation of these issues on the timing and budget of the project.

The engineering and fabrication of the joints for diagrid structures are more complex than for an orthogonal structure, thereby incurring additional costs. The precision of the geometry of the connection nodes is critical, so it is advantageous to maximize shop fabrication to reduce difficulties associated with on-site work and the erection of the odd geometries that are associated with the design of the diagrids.

The geometry and the tendency toward the use of prefabrication have necessitated a changed approach in the construction of diagrids, which must address:

- How is constructing a diagrid different from other structural types?
- How do choices in node design, member type and length as well as module size impact construction and erection?
- Transportation issues associated with nodes and long members
- Site issues that are unique to diagrid construction
- Stability during erection (weighing size of member versus temporary shoring)

Generally speaking, diagrids are more difficult to construct than standard structural steel buildings and this must be reflected in the way that diagrid structures are designed to be constructed to ensure a quality product and safety for the ironworkers.

Ironworkers are aligning a section of the diagrid basket column for the new entry pavilion at the World Financial Center in New York City, NY, USA, designed by Pelli Clarke Pelli. The all-welded construction required a very high level of fit to establish continuity in the curved steel tubes. The project was erected by Metropolitan Walters of New York and fabricated by Walters Inc. of Hamilton, ON, Canada.

SAFETY ISSUES

Awareness of the importance of safety on the construction site has become a significant issue. Ironworkers walking the steel without fall protection equipment, common practice in the assembling of skyscrapers in New York City in the 1930s, is no longer acceptable practice. The construction site must be maintained as a safe environment. Ironwork is a dangerous task. Erecting the skeleton of a structure with pieces of steel weighing many tons requires very careful planning and a highly skilled labor force that must function as a coordinated team.

Brookfield Multiplex Australasia, the owners of One Shelley Street in Sydney, are working with Professor Dennis Else (Group General Manager Sustainability, Safety and Health) on a working method called the "Safer by Design Strategy". The "Safer by Design Strategy" is based on growing a mature safety culture driven by demonstrated senior management actions. The senior managers have been encouraged to lead by role-modeling an approach where safety is embedded in the business processes and decision-making. This strategy has four key themes:

- being safer by design and planning
- focusing priority on critical risks
- demonstrating practices not just paperwork
- growing a mature safety culture

The statistical driver for this strategy rests in the high injury and death rate for ironworkers on construction sites, particularly as compared to general construction workers and the manufacturing sector. In spite of fall arrest protection, working extensively at great heights - before floors are constructed - puts the ironworkers at risk. According to "Safer by Design", 2005-06 statistics cite the deaths per 100 million hours worked as two for the manufacturing sector, four for the construction sector and 12 for ironworkers.

On the One Shelley Street project, the design strategy of the diagrid structure promoted the extensive use of prefabrication as a safety issue. This meant that 50,000 hours of work that would have been done at height on site was done off site in a factory or shop environment. The shop environment puts the workers "on the floor" or allows them to use short scissor lifts to access the steel. The shop will have cranes or other mechanisms that can rotate and move the steel elements so that the pieces are in the best position for access. This is particularly important for welding and grinding operations. This left only 3,500 hours of on-site rigging that was done from within scissor lifts or from the scaffold, which resulted in a 75% reduction in the risk of fatality.

Safety by design and planning underlies many of the strategic decisions on a complex project. Proper sequencing and maximized shop fabrication works in the best interest of all aspects of the project. Safety and access to the work on site is critical in the assessment of the constructability of the project and needs to be addressed during the design phase. This feeds into the nature of the prefabrication of elements and the maximization of aggregated element size in order to minimize the number and nature of on-site connections. The preference to make welded connections in the shop and bolted connections on site addresses safety issues as well as the aim of making the project easier to construct.

Where on-site welding is required, temporary bolted connections will allow the release of the crane. Costs increase due to the need to provide safe, temporary platforms from which to access the connections to complete the welding processes. Platforms and welding enclosures will vary by project, position of the

One Shelley Street in Sydney, Australia, designed by Fitzpatrick + Partners with Arup, was designed with constructability and safety as a premier concern. The part of prefabrication work was maximized. The majority of diagrid connections were made by bolting. The amount of work done at height was minimized.

The X-nodes and two diagrid elements for One Shelley Street were preassembled in the staging area, making the erection simpler and limiting the work done at height. Steelwork was completed by Bluescope.

connection to be welded and by climate/time of year. Practices will also reflect local construction and safety codes.

The ability to preassemble the node and diagrid elements on site is also effective in minimizing the work that needs to be done at height. It is also much quicker and easier for the ironworkers to access the pieces while they are sitting in the staging area to complete these bolted connections, something which is often required because of limitations on the transportable sizes of individual pieces.

The fully welded connections on the large tubular sections for the International Finance Center in Guangzhou, China, required the construction of scaffolding or platforms at each weld to provide safe access for the ironworkers. A very high-quality architecturally exposed finish required this sort of approach. The platforms must be attached to the diagrid members in such a way as to not damage the finish of the steel.

The diagrid-to-node connections for the Bow Encana Tower in Calgary, AB, Canada, designed by Foster + Partners, were welded to a high AESS standard (AESS4 by Canadian standards). As the project proceeded with construction through very harsh winter months, enclosed platforms were constructed around the nodes to protect the welders from weather exposure as well as the risk of falling.

One of the ground floor nodes for the Bow Encana Tower. One of the benefits of shop fabrication is the ease of access to the steel parts for welding. Steel must be preheated prior to welding. This is much more difficult on the site for reasons of access, weather and safety. The shop provides better quality control and a safer environment for the work.

ARCHITECTURALLY EXPOSED VERSUS CONCEALED STEEL

Whether or not the structure and its elements will be Architecturally Exposed Structural Steel (AESS) or concealed steel will have a huge impact on fabrication, handling and erection. Although the precision of fit and tight tolerances may be the same for AESS and concealed diagrid systems, resulting from the complexity of the geometry of the individual project, the handling of AESS must be much more careful. Where standard structural steel can be handled with devices such as "come-alongs" or heavy chains to urge fit, AESS will need to use padded slings at lifting points and will be damaged if undue pressure is required to forcefully fit the components. Typically most AESS will be either shop-primed or even shop-primed and painted, as it is easier to control the quality of the painted finish or intumescent application if applied in the shop. Even if the AESS arrives at the site without its final finish, the use of extreme pressure will damage the surfaces and require on-site remediation (which could include grinding and filling). While this is done on occasion, it is not desired.

For on-site connections that are welded, the finish must terminate at a distance from the joint to maintain the bare steel for welding operations. This means that after grinding and finishing operations are completed the steel will be primed and receive its final finish. If diagrid elements that are to be welded are designed as AESS, more care will be required in completing and finishing the welded joints. This will include the complete removal of all of the temporary tabs that are used to bolt the elements prior to welding.

If AESS is to be used, decisions need to be made regarding its classification of visual exposure at the outset of the project. This will impact the connection design and choice between bolting (easier on site) or welding (much more difficult and time-consuming on site). The scale of the project and the distance to view needs to be taken into account when deciding about the weld remediation that is to be done on site. If the distance to view is greater than 6m/18ft, the viewer will be unable to recognize extensive remediation. If the distance to view is less than 6m/18ft, it may make sense to more thoroughly remediate the welds or welded connections. This will have a significant impact on the cost of the project and the work carried out on site.

It is important to recognize the variation in global preferences for steel shapes and fabrication. This will impact the choice of member types for the project, which in turn also impacts the fabrication and erection process. Where Western projects tend to make use of hot-rolled wide flange or Universal sections for concealed structural steel, it is very common in Asia and the Middle East to use custom-fabricated hollow square or round sections for concealed structures. Due to the differences in labor costs and practices and predominance of the use of concrete-filled steel tubes these tend to be erected as fully welded structures. This is the normal practice on standard framed buildings and seems to be adopted fairly widely for other types of projects, including diagrids.

The Beijing National Stadium (Bird's Nest) in Beijing, China, designed by Herzog & de Meuron/Ai Weiwei/Arup, used custom-fabricated square sections that were welded on site. The temporary bolting tabs were removed and ground. Although traces of the connections are visible in some lighting conditions, the scale of the project is such that the final product is more than sufficient in quality.

The Chow Tai Fook Tower in Guangzhou, China, uses custom-fabricated hollow sections. These are joined using welded connections. As the structure will be concealed, the temporary tabs have been cut away and primed, but will not be further remediated.

ECONOMY THROUGH PREFABRICATION AND REPETITION

As the diagrid type of structure is more expensive to fabricate, cost and time savings can be realized by means to a high degree of repetition in the design and prefabrication of the nodes. A shop environment creates a safer workplace as well as one that can facilitate quality work. Given the high-profile nature of most diagrid buildings, pure repetition does not seem to be a priority as this would call for regular geometry in plan as well as uniformity over the height of the building. The Hearst Magazine Tower in New York City is one of the few diagrid structures with a simple rectangular plan and also with a uniform plan for all floors, resulting in a limited number of node types. Its design reflects the state of steel design and fabrication software of the time. Much has changed in the last 10 years. The development of new detailing software based on Building Information Modeling (BIM) has greatly changed the fabrication industry. The available software is capable of dealing with highly differentiated geometries and also of producing shop drawings for each element. If the fabrication method and surface treatments are relatively uniform, it is not difficult to customize the nodes.

For the Addition to the Royal Ontario Museum in Toronto, ON, Canada, designed by Daniel Libeskind, every piece of steel was unique. This meant that the handling and erection of each piece was different as the lifting points varied. The steel itself did not require overly careful handling, as this was a concealed structure, so if multiple attempts were required to fit the pieces, this resulted in time delays but not necessarily damage to the finish of the steel. Had this been an AESS project, its complicated geometry would have greatly increased the time and cost of the project. Steel fabrication and erection by Walters Inc.

Current digital technology has made a high level of differentiated prefabrication possible. Where the early diagrid structures such as Swiss Re and Hearst had quite regular plans and geometries that limited the number of unique variations of the nodes, projects such as Capital Gate boast having no two nodes identical. Jeff Schofield, one of the designers for Capital Gate in Abu Dhabi, describes its complexity as being the direct result of its “place and time”. The software and technologies were capable of fabricating what the client desired: a unique and iconic building. This was more important than the cost of such non-repetitive fabrication. In any case, the interoperability of the BIM software provided a high level of detailed communication between the engineer, the steel fabricator/detailer and the façade engineers.

If budgets are more modest, the overall design intentions should work toward minimizing the numbers of variations of the nodes and connections. Consistency in prefabrication will carry through to consistency in connection design. This will positively impact the erection schedule, as the ironworkers will be able to create some adeptness in dealing with angular connection geometries. Lifting points and lifting angles will become consistent and therefore much easier to deal with.

IMPACT OF NODE AND MODULE CHOICES ON ERECTION

There are two schools of thought as to the rigidity of the construction of the nodes themselves. In theory, if designing a simple triangulated structure with purely axial loads, in the logic of a truss, the center of the node need not be rigid and can be constructed as a hinge or pin connection, as it is transferring axial compressive or tensile forces. Where this may work in theory, it would require temporary shoring to maintain the diagrid members in position until the triangulation was physically complete. Even for symmetrical structures with well-balanced loads, and more so for eccentrically loaded structures, rigidity in the node is essential to assist in self-support during the construction process. In many of the diagrid projects constructed to date, the nodes have been prefabricated in the shop as rigid elements, allowing for incoming straight members to be either bolted or welded on site more easily and without need of temporary supports until the next node is attached.

Erection can be improved on an all-welded AESS diagrid if bolting can be used for some of the concealed connections. In the diagrid structure for Capital Gate the ring beams can be seen to be bolted to the nodes as they will be concealed. The greyish color on the diagonals indicates the area used to make the welded connection. This area is at a distance from the centroid of the node in order to make the welding work easier.

The use of a rigid node moves the connection points away from the centroid of the node, often to points that may be more than a meter away. This allows for easier access to complete the connections on the construction site, as it gives the ironworker more area in which to work. From a practical perspective, if using bolting, more than one ironworker can assist in completing the work at the node in cases where the connecting points are widely spaced apart from each other.

If the structure is to be clad or concealed, as in the case of the Hearst Magazine Tower, the diagrid elements can be bolted on site for speedier erection. In cases where the diagrid is designed as architecturally exposed, the connections may need to be welded. This adds significantly to the cost of erection, as more durable scaffolding is required for welders to access the nodes. It is more difficult to get high-quality on-site welds due to limitations of the access angles. A mixture of welding and bolting is an option. For Capital Gate, the connections of the diagonal members to the nodes that were to be architecturally exposed were welded, and the connection of the node to the ring beam at the edge of the floor that would be concealed was bolted.

Although the diagonal members on the Guangzhou IFC span four floors from node connection to node connection, they are braced by each of the floors. This assists in constructability and also allows for the welded connections to be less apparent. The node-to-member welds at the lobby level, where these connections are visible to the public, were more demanding.

Even where large modules are used, as in the Guangzhou IFC and the Bow Encana Tower, the diagonals are fabricated as a unit for added stability. The diagonals are not spliced on site. In the case of the Guangzhou IFC, where node-to-node spans are four floors, the extreme size of the members ranging up to 2m/6.6ft in diameter as well as the overall two-storey height of the node itself precluded preassembly of the node to diagonals on site. The nodes and diagonals were lifted separately. The use of the intermediate floor edge beams to brace the diagonals assisted in providing fairly immediate permanent support for the diagonal members. This decreased the need for temporary support.

TRANSPORTATION ISSUES

The transportation issues for diagrid structures are similar to those for more standard steel buildings. The size of the elements to be shipped needs to be carefully calculated in order to ensure clear passage from the fabrication shop to the site. Height and width restrictions on roadways and bridges, including the height of overpasses, must be known and adhered to. Depending on the precise location of the site, street widths and turning radii will also need to be known. This may result in significant shipping distances to transport the fabricated pieces to the construction site. It is essential to use a very experienced and qualified steel fabricator on complex diagrid projects.

Transportation limits feed back down into the maximum sizes of the elements to be transported. This will begin to determine where the connections for larger pieces need to be made, which in turn will indicate how the connections are to be done. The diagrid basket columns at the new entry pavilion to the World Financial Center in New York City had to be designed in a number of pieces in order to transport the elements. This made for a fairly complicated site assembly. The baskets were assembled in the shop to ensure that the fit was correct, then taken apart again for shipping. As the baskets increase in size from the base to the top, it was necessary to divide the top "ring" into six pieces. As the basket ribs are overlapping in a kind of woven pattern, the connections could not be clean. Additional X-shaped elements were needed as infill to complete the shape. (see p. 59)

Special trucks for extremely oversized pieces use the steel element to bridge between detached driving and rear ends. This allows the truck to turn corners as it will articulate. However, this method requires that the elements are designed not to deflect or suffer permanent distortion as a result of transport. This is an important issue for diagrid members, which are typically designed to withstand axial compressive and tensile forces but not bending. These sorts of vehicles will often require an official escort and have their travel times limited on public highways.

The base element for the columns of the new entry pavilion of the World Financial Center in New York City, NY, USA, designed by Pelli Clarke Pelli, (right) was shipped in one piece. The curved elements attached to the base are temporary; they allow for the rotation of the piece to a vertical position in order to lift it without damaging the base plate.

The logistics and expense of trucking the steel needs to be considered. This is one of the diagrid elements for the Bow Encana Tower (left). It had to be shipped from Hamilton, ON, to Calgary, AB, a distance of 3,340km/2,075mi. Only one element fit on the trailer. It was fabricated in one piece so that it did not need to be spliced on site and could be lifted in one piece. This meant that the welds would occur only at the node connections. Walters Inc., the steel fabricator for the Bow Encana Tower and the World Financial Center Pavilion, has developed an expertise in the design of oversized structures for transportation and erection.

SITE ISSUES

Providing the ironworkers with the proper access to the steel to finish the connections, particularly in the case of welded connections, is a significant site issue. Where low-level work may be completed from a lift truck, high-level work requires the construction of secure platforms. The scaffolding required to create platforms may be quite dense and preclude access to the site for other trades. This is often the case for the high-level steel in atrium spaces that is very difficult to reach and will often require the construction of significant scaffolding. The large diagrid columns (diagrids without an internal floor, acting like columns) that were created to support the roof of the new entry pavilion for the World Financial Center in New York City required significant scaffolding to complete the intense welding. It is important to ensure that this kind of scaffolding is budgeted for during the construction bidding process.

The construction of the diagrid columns for the new entry pavilion of the World Financial Center in New York City had very tight site conditions. The use of lift trucks was limited by the load capacity of the small staging area that was itself a cast concrete floor over the entrance to the subway below. The numerous welded connections required intensive preparation and remediation to prepare for painting. This in turn required the construction of a multiple-floor scaffold to provide the ironworkers with good access to complete the work. The on-site steelwork was completed by Metropolitan Walters.

The Chow Tai Fook Tower in Guangzhou, China, makes extensive use of welding. This requires the construction of many platforms to provide access to the connections.

It is also important to be able to provide an adequate staging area at the site for the preassembly of components. Depending on the size of the members and transportation restrictions, it may benefit the erection sequence to create larger assemblies for lifting on the site. This may include lifting nodes with some members attached, which was routinely done on the the Swiss Re Tower (see p. 40) as well as on One Shelley Street (see p. 112). The best configuration for erecting assemblages of a node with diagonals is an inverted V. This configuration limits the deflections on the long members, makes fitting easier and minimizes the need for temporary support or bracing.

Ironworkers on the Royal Ontario Museum are preassembling members prior to erection. This procedure was critical to the success of the project because of the highly heterogeneous nature of the assemblies and lifts. Combinations of nodes and diagrid members would have been too large to be transported to the site as units.

The elements of the ArcelorMittal Orbit Tower, designed by Anish Kapoor and Cecil Balmond with Arup for the London Olympics in 2012, were prefabricated and prepainted in the shop. The star-shaped assemblages of node and diagrid pieces were sub-assembled on site and lifted in aggregated pieces to save time. While the welding of the node elements was done in the shop, on-site connections were predominantly bolted. The finish on the bolts is actually bright and highlighted as a design feature. The steel was fabricated and erected by Watson Steel Structures.

MAINTAINING STABILITY DURING ERECTION

As most diagrid buildings tend to be constructed with steel as the material used for the perimeter diagrid and for the floor structure, it is extremely important to consider that an all-steel building that does not use reinforced concrete for additional lateral stability as in a core, will not be fully stable until the concrete floors are poured and provide diaphragm action. Even if using a reinforced concrete core for lateral load resistance, the diagrid members will not be fully self-supporting until they are triangulated and the connections completed.

A temporary shoring and stabilization system was used on the Manukau Institute of Technology in Auckland, New Zealand, designed by Warren and Mahoney Architects. The completed floor system used precast concrete Double Tees and steel decking in combination with a significant reinforced concrete topping. The temporary stability system had to remain intact until the overall structure was complete. The location in a high seismic zone impacts the design of any temporary stability systems.

The members-and-node connections for the Royal Ontario Museum were designed to be essentially self-supporting. No shoring towers were required. Stabilization was achieved using tension cables until the floors were completed. These assemblages present trip hazards on the site but do not create as much interference or cost as more massive shoring.

The temporary support system for the floor beams on One Shelley Street needed to stay in place until the floor system was completed and able to stabilize the structure.

This is one of the main reasons why the nodes of a diagrid construction are typically fabricated as stiff. On tower buildings, the preassembled node and diagrid elements can often be installed in an inverted-V-configuration in order to minimize the cantilevering effect on the diagrid elements. If the diagrid elements deflected before being joined to the next node, this would make the erection difficult.

Where temporary shoring is required, it is important to understand its impact on traffic flow on the construction site. It can be anything from a trip hazard to a complete obstruction to work.

Constructability was part of the rationale behind using a diagrid for the CCTV Building in Beijing, China, designed by OMA with engineering by Arup. The diagrid was essential in maintaining the stability of the cantilevered sections prior to their connection. Significant welding work had to be done at height, requiring the construction of scaffolding and enclosures at the welds. The timing of the final connection of the two cantilevered sections was based on the documented movements of the separate towers to reflect both deflection and the impact of temperature. The relative movements of the towers during the day were found to be around +/-10mm (+/-0.39in). The contractor made the final measurements of the gap exactly 24 hours beforehand (i.e. at identical ambient conditions) so that final adjustments could be made to the lengths of the seven linking elements while they were still on the ground. The elements were lifted into place - to less than 10mm tolerance - and temporarily fixed with pins before the towers started to move relative to each other. The pins allowed them to carry the thermal loads while the joints were fully welded over the following 48 hours.[1]

9 FAÇADE DESIGN

The International Finance Center in Guangzhou, China, designed by Wilkinson Eyre Architects, uses a rectilinear curtain wall pattern in combination with relatively transparent glazing to reveal the diagrid structure of the building.

CURTAIN WALL AND FAÇADE DESIGN

Although the use of a diagrid is not always associated with odd geometries and forms, many diagrid buildings purposefully choose this structural system in conjunction with a desire to create high-profile buildings that use geometry and form as part of their aesthetic. This will often require the use of non-standard curtain wall systems, particularly in the case of curved or faceted geometries. Yet although diagrid structures themselves are based upon triangulated geometries, not all diagrid buildings translate this directly into the façade treatment.

Façade design needs to address the following concerns:

- → geometry of the building
  - → scale and size of the building
  - → size of the module
  - → planimetric shape of the building
  - → curvature
- → placement of the structural diagrid inside or outside the envelope, impacting physical engagement with the curtain wall
- → expression of the diagrid
  - → diagrid emphasized in curtain wall design and mullion placement
  - → curtain wall in front of structure and diagrid not directly expressed in the cladding
- → use of space/function of the building
  - → open floor area (mullions need not coordinate with abutting partition walls)
  - → partitions abutting glazing (issues of coordination between diagrid, mullions and partition walls)
- → budget for curtain wall
  - → size of units
  - → type of glass (number of panes)
- → single or double façade
- → shading strategies (especially coordination of blinds and window coverings with triangular schemes)
- → issues of natural ventilation
- → rectilinear versus triangular system as the desired aesthetic
- → cleaning and maintenance considerations

These parameters lead to a choice between a triangular and a rectilinear pattern in the curtain wall system as well as a variety of methods of expression of the modularity of the diagrid on the façade. Although diagrids are a structural choice, this choice is often motivated by the quest for a certain look or aesthetic, hence the current tendency in design to express the diagrid on the building to showcase its use.

The decision whether or not to emphasize the diagrid frame in the façade and the size of the diagrid in the cladding is primarily an aesthetic choice. The choice to express the location of the structural diagrid in the curtain wall varies from project to project. This is driven by the architecture and not the engineering.

The decision to use a triangular versus a rectilinear curtain wall system will take into consideration the overall size of the diagrid structure as well as the form of the building itself. The modularity of the curtain wall will usually scale down the dimensions of the diamonds or triangular shapes to suit the height of the floors and the requirements for fixed and operable windows. Buildings with more curvilinear forms tend to use a triangular curtain wall that adapts more easily to the form. Diagrids with large modules can more easily accommodate standard recti-linear curtain wall systems as infill and are therefore easier to "fit" on the façade. Choices will also reflect the desire to express the diamond grid - in cases where the horizontal structural components are purposefully not expressed - versus a triangular pattern that expresses the placement of the horizontal tie beams.

TRIANGULAR GLAZING

Many diagrid buildings, in particular those with a curvilinear form, tend to subdivide the diagrid into triangular glazing. Triangular geometry creates something like a mesh to allow for the approximation of the curved form through straight elements. This is far more economical than using genuine curves, even if the glazing system itself is more expensive than a traditional curtain wall system. However, the use of curved forms can take advantage of some interoperability among design and detailing software that is used by the architect, structural engineer, steel fabricator and façade engineer. The development of façade engineering as a consultancy over the past decade or so has resulted from an increasing performance demand for envelopes regarding energy concerns and and also from changes in the form of buildings, many of which belie the use of traditional curtain wall systems due to their odd geometries.

The scale of the faceting must be appropriate to make the building appear curved despite the absence of bent elements. The specific size or subdivision of the triangulation of the structural diagrid into glazed (or opaque) elements will also depend on whether all glazed units are fixed panels or whether operable units are to be incorporated into the façade. This may be a function of the use of the building. Natural ventilation is often required for residential occupancies, including hotels. Operable windows have been incorporated into a number of diagrid buildings using triangular windows. Awning types with the point down seem to be the dominant choice, as these permit air flow but are oriented to prevent rain infiltration.

A triangular mesh seems to be draped over the form of the Guangzhou Opera House by Zaha Hadid. Triangular forms are well suited to conforming to such curved geometries.

The façade design for the Swiss Re Tower (30 St. Mary Axe) in London, England, designed by Foster + Partners with Arup, makes simple use of the mullion color and the transparency of the glazing to express the diagrid. The façade uses a mix of diamond and triangular-shaped glass, the latter allowing for some operable windows.

In the Aldar Headquarters in Abu Dhabi, the steel diagrid is covered with gypsum board in the interior of the building and by the building envelope on the exterior. Yet the design made the aesthetic choice to express the location of the diagrid elements on the exterior façade, creating a prominent diamond pattern that clearly presents the diagrid module as an element of design. The expression of the eight-storey module on the façade creates a highly visible pattern.

The diagrid on Aldar Headquarters in Abu Dhabi, UAE, designed by MZ Architects, is expressed through the use of an accentuated diamond-shaped element on the cladding system.

The installation of the curtain wall system and the column covers at Aldar Headquarters.

The triangular pattern within each of the large diamond modules is independent of the adjacent modules. This makes it possible that the horizontal mullions do not align. The open plan of the office space avoids any issues with the coordination of partition walls. The reflective glass was chosen in part to reduce solar gain.

The design for Capital Gate in Abu Dhabi makes a more unobtrusive gesture on the façade regarding the location of the diagrid situated behind the triangular glazing. While the two-storey module size, designed to support the building's 18° lean, constitutes one of the smallest modules to date, the member sizes are large and to translate this to the visual appearance of the façade would have been quite overbearing. Instead, a slight color change at the visible grid is used to acknowledge the pattern of the diagrid structure. In general, these sorts of twisted building forms tend to subdue the reading of the module through the façade.

The façade system for Capital Gate was prefabricated as diamond-shaped units measuring about 8m x 8m/26ft x 26ft to match the diagrid structure. As each of these units was unique, a specialized extension of the BIM system used for the structure was employed to design the façade. The panels are braced at mid-height by the floor for wind resistance. Operable units are incorporated into the double façade system of the hotel. The façade of Capital Gate was engineered by Hyder Consulting and constructed by Waagner Biro.

The glazing panels of Capital Gate are framed in a prefabricated steel lattice that sits away from the larger diagrid member as a way to make the building appear more smoothly curved. A structural steel element braces the façade system back to one of the ring beams.

The mullion system of Capital Gate (left) assumes a very low profile. The silver pressure caps are located at the diagrid members. An LED lighting system is integrated into the façade. The lights are located at the intersections of the mullions.

The prefabricated steel framing for the glazing system of Capital Gate (right) is attached to the diagrid at the nodes. The mechanics of the attachment system is discreetly hidden from view.

RECTILINEAR GLAZING

A rectilinear curtain wall is often chosen for projects whose forms are relatively planar. It is typically less expensive than the customized triangular option. For commercial office functions, these windows provide a practical solution for the provision of blinds and shades and also for partition walls, if a subdivision of the interior needs to extend to the perimeter of the building. Many façade solutions for tall diagrid buildings, like the Guangzhou IFC and the Bow Encana Tower, employ windows that are of the same height as the floor-to-floor distance. Larger glazing units are more economical than systems comprised of smaller units. The visual scale of a building can be manipulated by the choice of window proportions.

The Hearst Magazine Tower in New York City, NY, USA, designed by Foster + Partners and engineered by WSP Cantor Seinuk, with its fairly large diagrid modules uses a rectilinear curtain wall glazing system within the triangular areas. This allows for the use of partitions, if desired, and standard window shading products. A large curtain wall cap system highlights the presence of the diagrid on the exterior.

A detailed view of the north façade of the Bow Encana Tower showing the profile of the column covers that have been used to express the diagrid over the rectilinear glazing pattern.

The Bow Encana Tower in Calgary, AB, Canada, designed by Foster + Partners with Zeidler Partnership, uses a system that is similar to the precedent set in the Hearst Magazine Tower. The curved south façade (top) is faceted rather than requiring more expensive bent treatment. As the space behind this façade is part of a double façade atrium system, there is no concern regarding the alignment of interior partitions.

The rear façade of the Bow Encana Tower (bottom) uses a similar curtain wall treatment but adapted to reflect the office use behind. The design language is consistent and expresses the structural pattern of the truss tube frame. The building scale again permits the use of faceting to achieve a curved appearance.

The Tornado Tower in Doha, Qatar, designed by CICO Consulting Architects and Engineers, uses an expressed system to highlight the diagonals, but not the horizontal bands of its diagrid structure, with LED lights installed at the nodal points for night lighting of the structure. A rectilinear glazing system has been used that runs continuously behind the diagonal bands.

This construction image of the Tornado Tower (left) reveals that the rectilinear curtain wall has been installed continuously and that the bright bands highlighting the placement of the diagrid behind have been installed in front of the curtain wall. At the right is the Doha Tower under construction, its diagrid concealed by the mashrabiya screen that is erected over the curtain wall to provide shading.

Another important factor for the selection or design of the curtain wall system is the frequency and nature of operable window units. Natural ventilation is seeing an increase, even in some high-rise types. Combined uses, such as office and hotel, might also infer differentiated glazing, both in terms of transparency as well as inclusion of operable windows.

The exterior diagrid at One Shelley Street in Sydney, Australia, designed by Fitzpatrick + Partners with Arup, permits the use of a rectilinear curtain wall system and standard window shades on the interior.

The rectilinear curtain wall of the Al Bahar Towers is installed continuously. The banding that highlights the location of the honeycomb structure sits proud of the curtain wall. It is fitted with an LED lighting system to showcase the structure of the towers at night.

The Guangzhou IFC uses a very transparent glass in combination with a rectilinear curtain wall. The massive tubular round diagonals are clearly visible through the glass and lit at night to enhance their expression. The dimensions of the curtain wall elements are modified at the rounded corners of the building to accommodate the tapering of the tower. A special ventilation system marks the refuge floors. The transparency of the glass was carefully considered to allow the diagrid to show while also controlling solar gain.

A different glazing system is used on the double-height ground floor lobby areas of the Guangzhou IFC to highlight the 2m/6.6ft diameter tubes. The glazing system is based on stainless cables and a slight triangulation of the glass panels.

CLEANING AND MAINTENANCE

Irregularly shaped diagrid buildings must consider cleaning and maintenance as part of the design problem. Standard window washing stages that use a track system that is integrated into the curtain wall will not work on buildings with inward and outward curves or those with odd inclines and irregularities. It is also important to consider climate and location-related issues as these will impact the required frequency of cleaning the building.

A customized cleaning track was fabricated to work with the 15° lean of the Puerta de Europa Towers in Madrid. The structural track was integrated into the cladding pattern that highlights the location of the diagonal grid.

A detail of the façade of the Bank of China Tower in Hong Kong, designed by I.M. Pei and engineered by the firm of Leslie E. Robertson Associates, showing the expression of the diagonals within the same plane as the rectilinear curtain wall patterning. This corner and flush façade detailing puts very stringent requirements on alignment and coordination. The vertical expanses of the curtain wall make use of relatively standard cleaning equipment. Operable windows have been incorporated into the curtain wall design.

To address the “gherkin” shape of the Swiss Re Tower in London, an extendible rotating crane was designed for the top of the tower. The crane runs on a parallel set of rails that are tied back to the building structure. The arm extension allows the accommodation of the differing diameter of the tower. The crane rails are designed to become a design feature of the top of the tower.

The cleaning stages for Swiss Re run on cables that are clipped to the curtain wall at select reinforced points over the height of the structure. Additional foam rings are used to prevent the cables from striking and damaging the glass and mullions of the tower. A crane-like arm extension is used to suspend the cleaning stages. A set of rails at the top of the tower is used to allow the arm to travel the circumference of the building.

Oddly shaped buildings may require the design and manufacture of specialty cleaning equipment. This was the case for the Aldar Headquarters in Abu Dhabi. The disk shape of the building lends itself to abseiling on the outwardly curved sides, but the narrow sides of the disk could not use this means. A special window washing lift was fabricated that runs on tracks and climbs up and over the building, providing access for cleaning. The machine is adjusted as it travels around the building.

This specialized piece of equipment is used to provide cleaning access for the glazing on the sides and top of the Aldar Headquarters. The device is designed to rotate in response to the curvature and runs along tracks on either side of the platform.

The CCTV Building in Beijing, China, designed by OMA with engineering by Arup, has a very different approach to the expression of the diagrid in the cladding system. Solid recessed panels highlight the locations of the braces. A rectilinear glazing system is used to allow for partitioning and standard window shade systems to be used. As the glazing system wraps the underside of the cantilevered section, a square grid results. The large cantilevered portions of the building requires a different approach to cleaning as they are beyond the reach of standard lift trucks. This is a view of the track system integrated in the underside glazing which is used to suspend window washing equipment.

A different option is cleaning the façade by abseiling, a form of rappelling down the façade. In the case of Capital Gate in Abu Dhabi, the requirements for cleaning the building were incorporated into the design at a very early stage. Abseiling is used to clean the glazed and mesh screen portions of the exterior. Hooks are integrated into the façade system so that the abseilers can manoeuvre themselves close enough for cleaning, especially on portions of the façade where the tendency would be to hang away from the glass. This method is very common in the humid and dusty environment of the UAE and for this geometry of building.

This view of Capital Gate during construction demonstrates the severity of the cleaning problems experienced in the UAE due to the humid, dusty environment. A maintenance plan was essential.

The exterior of Capital Gate in Abu Dhabi, UAE, designed by RMJM, being cleaned by a team of abseiling window washers.

The portion of the Capital Gate façade that is situated between the sun shade and the curtain wall is accessed by abseiling for cleaning.

The bottom line with respect to the selection and design of the curtain wall system is that it is greatly impacted by the dimensions of the base module for the diagrid in combination with the geometry of the building. Larger modules allow more flexibility in the choice of curtain wall. Smaller modules are more restrictive and invariably lead to more complex glazing systems that tend to be triangular. This impacts labor costs, as complex systems tend to be more time-consuming to install and maintain as they cannot be fitted with standard window washing equipment.

10

EXTERIOR DIAGRIDS AND DOUBLE FAÇADE SYSTEMS

The placement of the diagrid outside of the thermal envelope at One Shelley Street in Sydney, Australia, designed by Fitzpatrick + Partners with Arup, makes the building quite unique. This kind of diagrid placement is only possible in a climate where thermal bridging is not necessarily a key issue.

EXTERIOR DIAGRIDS

The use of exterior bracing has a rich history in High Tech architecture and Architecturally Exposed Structural Steel (see Chapters 5 and 6 in *Understanding Steel Design*), so it is not surprising to see temperate climates combine this precedent with the use of a diagrid structure. Where external bracing has been used in High Tech or AESS applications as a supplemental system of lateral reinforcement, exterior diagrids are acting as the primary means to satisfy the lateral loading resistance.

The Neo Bankside housing project in London, England, designed by Rogers Stirk Harbour + Partners, shows the use of exterior AESS bracing in conjunction with a non-diagrid primary structural system. This bracing system is using elliptical tubes. The bracing design must contend with issues of corrosion protection, differential thermal expansion and connection back to the primary structure.

The majority of structural steel diagrids have been placed on the interior of the envelope. This is especially critical in cold climates where thermal expansion is significant and thermal bridges must be avoided under all circumstances. In highly corrosive environments, exterior diagrids can also require ongoing maintenance due to oxidation and weathering of the finish. However, diagrids have been successfully used outside of the envelope in several instances, more frequently in hot or temperate climates, and to satisfy different programmatic or design motivations.

On the Manukau Institute of Technology in Auckland, New Zealand, designed by Warren and Mahoney Architects, the workers are installing the shading support frames that will sit between the exterior diagrid and the glazing system. The low five-storey height of the building and the moderate climate minimize issues related to thermal expansion.

One Shelley Street, Sydney, Australia

Perhaps the most notable exterior diagrid structure to date is found on One Shelley Street. The temperate climate has permitted the diagrid to be placed outside of the curtain wall façade, effectively maximizing the internal leasable area. Although the designers were originally planning to construct the members from architecturally exposed steel, it was decided for economic reasons to instead clad conventional Universal sections that were able to be hot-dip galvanized. This simplified the detailing of the diagrid and allowed for bolted site connections. As corrosion resistance is a great concern in Australia, this was likely a prudent decision. The exterior does not show the indispensable triangulation of the diagrid, as the required horizontal element is completed by the floor edge beams, which are situated on the interior of the curtain wall and out of view. The use of a rectilinear glazing pattern also serves to conceal the triangulation.

The steel sections for One Shelley Street in Sydney, Australia, designed by Fitzpatrick + Partners with Arup, were all hot-dip galvanized. This level of corrosion protection is crucial in a marine environment. As the sections were to be clad, it was also important to provide this extra level of protection to prevent hidden corrosion.

The innovative free-floating corners of the diagrid were part of the design motivation that drove this unusual detailing.

The exterior galvanized diagrid is connected back to the floor edge beam. This has been left untreated as it will be located on the interior of the building envelope. The penetrations to the envelope must be carefully detailed to prevent moisture intrusion.

The connection between the diagrid and the façade required custom detailing of the curtain wall at the points of penetration. The distance between the diagrid and the curtain wall must be sufficient for window cleaning. This type of structure will negate the use of standard window washing equipment.

The exterior diagrid terminates above the level of the green roof. The roof is not accessible, so the diagrid is not required to act as a fall barrier.

Where the exterior diagrid is used to enclose the double-height exterior space at the entrance, round HSS members are employed to brace the nodes back to the floor structure above, unlike the situation for other nodes that frame back directly to the floor system.

The exterior diagrid comes down to grade in a freestanding fashion and allows for the simple enclosure of the entry areas to the complex.

Canton Tower, Guangzhou, China

The external, concrete-filled tube diagrid structure for the Canton Tower creates a dynamic statement. Its function as a broadcast and sightseeing tower, comprising a wide range of heterogeneous interior spaces and some open portions, does not require a continuous enclosure of the floors and allows the diagrid to be clearly expressed as a separate element. The exterior use of concrete-filled tubes is appropriate in this relatively temperate climate. It is important with this degree of exposure to ensure that the coating system on the exposed steel is durable. The structure itself requires regular maintenance, inspection and cleaning.

The spiraling diagonal system required careful planning to coordinate the connecting elements to the floors of the enclosed portions. The use of advanced software has been essential in creating the individualized shop drawings for the fabrication of all of these elements.

As a structural system, the exterior diagrid of the Canton Tower in Guangzhou, China, designed by IBA with Arup, is clearly separated from the discontinuous spaces of the tower.

This view out from the observation floor shows the spiraling members of the diagrid structure as they pass the windows. They act as the horizontal elements of the triangulation. The façade system is independent of the diagrid pattern. By contrast, the floor-supporting connections occur on the more vertical diagrid members. These are larger to transfer the loads and make the load paths more direct.

The exterior location of the diagrid structure requires the clear functional separation of the diagrid from the floor system that supports the enclosed portions of the tower. Unlike other diagrid structures that use the floor system to triangulate the structure, the Canton Tower has an additional ring of spiraling elements to complete this function. One reason for this solution is the discontinuous nature of the sections of the building: there are open portions where support and triangulation could not have been provided by the floor system.

The beam element connects the floor structure of the observation level to the vertical diagrid member. The welded connections on the tube have been neatly done and not ground.

A view of a connection between the floor framing system and the exterior diagrid during construction. The round geometry of the circular concrete-filled tubes matches the spiraling dynamic of this diagrid structure. Rectangular custom-fabricated steel sections support the floors. A pin-type connector has been designed to make the physical and geometric transition between the two systems.

O-14, Dubai, UAE

This unusual building, a 22-storey-tall commercial tower atop a two-storey podium, employs a reinforced concrete exterior support system whose structural action resembles a perimeter diagrid system in that it is assuming the majority of the lateral loading. With O-14 the designers wanted to optimize the free-spanning space and create an exterior that, to a certain extent, serves as structure, skin and a shading device. Although this concrete exoskeleton is not a typical diagrid in the strict terms of the use of nodes and members of a steel system, the concrete exoskeleton is the primary vertical and lateral structure for the building. It frees the core from the burden of lateral forces and helps create highly efficient, column-free open spaces. The core, which is traditionally enlarged to receive lateral loading in most curtain wall office towers, was minimized to handle only some vertical loading, utilities and transportation. Also, while the typical curtain wall tower configuration results in floor plates that must be thickened to carry lateral loads to the core, the floor plates were minimized to only respond to span and vibration.

The overall view of O-14 in the Business Bay district of Dubai, UAE, designed by RUR Architecture (Reiser + Umemoto) with engineering by Ysrael A. Seinuk, illustrates the variation in the perforated exterior reinforced concrete shell.

The placement of the steel reinforcing bars in the perimeter wall follows a diagonal pattern which leads the load path around the openings.

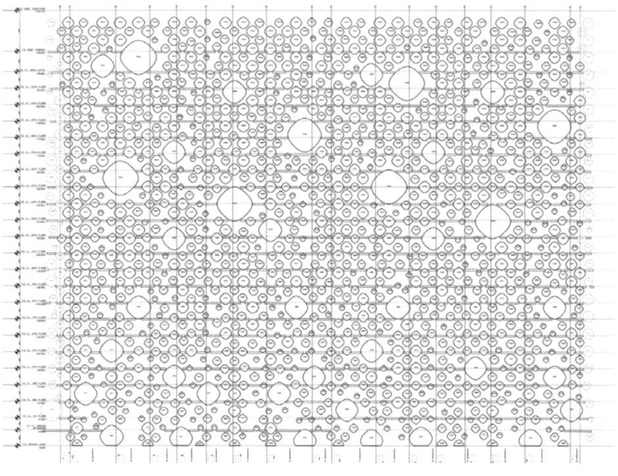

This drawing of the "unrolled" exoskeleton was prepared to assist with the layout of the openings and coordinate their relationship to the floor levels and floor support system.

The main shell is organized as a diagrid, the efficiency of which is wed to a system of continuous variation of openings, always maintaining a minimum structural distance between adjacent openings, adding material locally where necessary and taking away where possible. This efficiency and modulation enables the shell to create a wide range of atmospheric and visual effects in the structure without changing the basic structural form, thereby allowing for systematic structural analysis and consistent construction. The project essentially creates a double façade wall system, based on the ability of the concrete diagrid exostructure to also provide shading to the office space inside.

Although the design might appear to be randomized, slip forming could be used to work the formwork throughout the height of the building.

A view up the tower from the podium building showing the penetration of the pedestrian bridge through one of the larger openings.

The exoskeleton provides solar shading from the intense Dubai heat while still allowing relatively unobstructed views of the Business Bay district.

DOUBLE FAÇADE APPLICATIONS

Double façade systems for building envelopes are seeing increasing use in many environmentally motivated designs. Advantages include:

- the reduction in energy consumption that is provided by the extra protection
- protection of shading devices
- buffering of urban noise
- potential for incorporation of protected natural ventilation

The double façade system will normally consist of a double-glazed, more traditional curtain wall with a single-glazed additional layer. It is most typical for the single layer of glazing to be on the outside, although this will vary as a function of the severity of the climate. The cavity itself can be ventilated or sealed.

Diagrid structural systems have been incorporated into double façade applications in a number of ways:

- The double façade cavity can be located behind the diagrid structure.
- The primary perimeter diagrid structure can support an additional exterior screening device, a solution that requires additional detailing for connections and specialized façade systems.
- A lighter-weight diagrid can be used outboard of a more standardized structural system and support the second skin, the outer layer of glazing.

In the latter instance, although not used structurally in terms of the support of the building, the buildings do take advantage of the structural attributes of diagrid design. These advantages include the inherent stability due to the triangulation of the diagrid, the diagrid aesthetic incorporated into the façade and the glazing design, as well as the effectiveness of prefabrication of the components.

The Bow Encana Tower uses the strength of the diagrid structure on its south façade to create extremely large atrium spaces in combination with a double façade. The diagrid structure has the unusual situation of being somewhat divorced from the structural support normally provided by the floor system due to the requirement to enclose a multi-storey double façade atrium space. While the exterior of the double façade is double-glazed because of the severity of the winter climate, single glazing separates the double façade cavities and atrium space from the office use. The interior side of the triangulated diagrids received a high-level AESS finish because of the visual exposure to the atrium, while the concealed backside could be finished more simply, saving cost.

Atrium space in the double façade portion of the Bow Encana Tower in Calgary, AB, Canada, designed by Foster + Partners with Zeidler Partnership. The view of the top floor (top) clearly shows the lack of structural support for the diagrid system. A view from a lower level (bottom) shows the limited nature of the connections between the office floor system and the diagrid structure.

A close view of the installation of the diagrid double façade system at the Cleveland Clinic in Abu Dhabi, UAE, designed by HDR Architecture. Special discreet bolted connections have been designed to make for both a detail and quick erection. The connection is recessed to preserve the clean visual lines of the white diagrid.
The system is braced back to the primary structure to complete the triangulation. This results in a two-storey diamond module. The diagrid is separated from the building to allow for cleaning access via the catwalk that also acts as a shading strategy. A fairly standard curtain wall system sits behind the outer double façade layer.

The Leadenhall Building, London, England

In The Leadenhall Building, a diagrid structure of a megaframe type has been integrated in a double façade system. The diagrid structure is situated between the inner and outer layers of glazing. In this case, the double-glazed layer is located between the diagrid structure and the interior office spaces, while single glazing is used on the exterior face. The single-glazed exterior layer provides buffering for temperature and noise as well as protection for the shading system. The external glazing incorporates vents at node levels to allow outside air to enter and discharge from the cavity. Controlled blinds in the cavity automatically adjust to limit unwanted solar gain and glare.

This rear view of the diagrid structure, to be comprised between the layers of the double façade, shows the special sliding connection with the floor support system that is required in order to accommodate the differential movement of the diagrid due to its location within the double façade cavity.

The grating system that aligns with each floor is fitted between the diagonals. The vertical supports for the exterior glazing have been installed. Workers are installing an exterior glass panel using a hoist system. The backwards lean of the façade makes this more difficult than installing glazing on a vertical face.

The horizontal beams on the diagrid of The Leadenhall Building in London, England, designed by Rogers Stirk Harbour + Partners with Arup, express the partitioning of the double façade system on the exterior. The top and bottom flanges of the beams extend out further than the diagonals, so that the glazing can bypass the vertical and diagonal elements. Venting occurs where the glazing meets the horizontal beams.

This exterior corner detail of the double façade system shows the relationship of the components. A tubular steel member runs up the corner to provide physical support for the very thin glazing mullions that support the single-glazed exterior.

Doha Tower, Doha, Qatar

The 45m/148ft diameter cylindrical volume of the Doha Tower is supported by a perimeter concrete-filled steel tube diagrid. One of the major design concepts was to use a double-skin system to control the sunlight in this hot desert climate. The floor framing system is cantilevered out to create a marked separation between the diagrid structure and the façade support system. The curtain wall systems follow a rectilinear pattern. The exterior skin is composed of four "butterfly" aluminum elements of different scales to evoke the complexity of the Islamic mashrabiya while serving as protection from the sun. The pattern and density of the exterior skin vary according to the orientation and respective needs for solar protection: 25% toward the north, 40% toward the south, 60% on the east and west. The internal layer is a slightly reflective glass skin that completes the solar protection. Additionally, a system of roller blinds can be used if needed.

The exterior shading system varies in density as a function of the orientation. The system is additive, with increased shading being provided through the addition of a finer layer of aluminum screen.

The Doha Tower in Qatar, designed by Ateliers Jean Nouvel with engineering by Terrell Group and China Design Construction International, separates the diagrid structure from the double façade support system, as is clearly visible in this view of one of the office floors.

Looking out through the mashrabiya.

This view on the top floor with the lighter-weight steel framing system of the dome shows how the exterior articulated skin is providing shade to the interior.

Photographs courtesy Ateliers Jean Nouvel ©CSCEC.

Al Bahar Towers, Abu Dhabi, UAE

The circular Al Bahar Towers use a honeycomb variation of the diagrid structural system. It is clearly visible on the curtain walls on the north façades of the buildings that have been designed without shading, as overheating due to direct solar gain is less of an issue for this orientation and the views to the city have been preserved.

The Al Bahar Towers have won several awards, including the CTBUH 2012 Innovation Award, for their light-responsive exterior façade system. The dynamic façade was conceived as a contemporary interpretation of the traditional Islamic mashrabiya, a vernacular form of wooden lattice screen used as a device for achieving privacy while reducing glare and solar gain. The mashrabiya at Al Bahar Towers is comprised of a series of semi-transparent, umbrella-like components that open and close in response to the sun's path. Each of the two towers includes over 1,000 individual shading devices that are controlled via the building management system to create an intelligent second façade.

The Al Bahar Towers in Abu Dhabi, UAE, designed by Aedas Architects with engineering by Arup, use a light-responsive mashrabiya shading system to create what is essentially a second façade on all but the north façade, which is left exposed.

This view of the mashrabiya panels in their fully open position (top left) was taken prior to commissioning of the building. The intelligent system results in highly uniform, automated opening and closing of the screen as a function of the time of day and light conditions.

When the mashrabiya is in its fully closed position (top right), the majority of the harsh desert sunlight is prevented from striking the glazed façade. Although the mashrabiya may appear very solid, it is created from a PTFE screen material to allow some light to the interior.

This view of the space between the screen and the curtain wall façade (left), taken close to completion of construction, demonstrates the nature of the transparency of the screen and shows the support system. The high level of dust and particles present in Abu Dhabi, combined with high humidity, presents maintenance issues for the façade. The interstitial space has been sized to provide access for cleaning.

According to Aedas Architects, each unit is comprised of a series of stretched PTFE (polytetrafluoroethylene) panels and is driven by a linear actuator that will progressively open and close in a daily sequence calculated to prevent direct sunlight from striking the façade and to limit direct solar gain to a maximum of 400 watts per linear meter. The entire installation is protected by a variety of sensors that will cause the units to open in the event of overcast conditions or high winds. The benefits of this system include: reduced glare, improved daylight penetration, less reliance on artificial lighting, and over 50% reduction in solar gain, which results in a projected reduction of CO_2 emissions by 1,750 tonnes per year.

Sky gardens, located along the southern façades between the curtain wall of the office spaces and the mashrabiya layer, alleviate the effects of solar exposure.

Capital Gate, Abu Dhabi, UAE

Iconic diagrid buildings located in the UAE may not be synonymous in most minds with sustainability, but sustainable design was an important consideration in the design of Capital Gate. While the fabrication and erection costs for such a complex structure might be higher, the organic, curved form is more aerodynamic, presenting less resistance to the wind and thereby reducing the respective structural requirements. The round perimeter was able to enclose the volume more efficiently than a rectangular floor plate of equal area. This saved both structural and façade materials and their embodied carbon costs.

In particular, there was a desire to use the façade to decrease the solar penetration in order to reduce the cooling load. An additional AESS and metal mesh screen, called the "splash", was used on the entrance side of the tower to cut down on solar gain into the office levels below. The stainless steel wire mesh is 90% open. The screen extends to become a large canopy that provides shade for the drop-off zone. The flexible system is designed to follow the curves of the tower. It is braced to the tower every five floors, with additional cross-bracing rods at every floor. The mesh eliminates approximately 30% of the heat that would reach the façade.

The underside of the "splash" as it flows away from the tower to become the shade for the entrance.

The metal mesh sun-shading screen is tied back to the nodal points of the diagrid with round HSS tubes.

The structural supports for the "splash" tie directly back to the diagrid structure through the cladding system.

Details of the supports for the "splash" system over the main entrance driveway.

The upper half of the tower, which houses the hotel, has a modified double-skin façade to reduce the solar gain. The system recycles interior air from the guest rooms into the façade cavity. This cavity creates an insulating buffer between the extreme exterior heat and the cool interior. The air is reused in the room rather than exhausted and replaced with 100% outside air. The low-emissivity glazing assists in cutting down heat gain through the exterior layer of the façade while maintaining a high degree of transparency.

This rendering of the double façade system at the hotel floors indicates the size of the spacing between the layers of glazing to permit access for cleaning. The curtain wall that separates the hotel rooms from the exterior façade uses a simpler, less expensive, rectilinear mullion system.

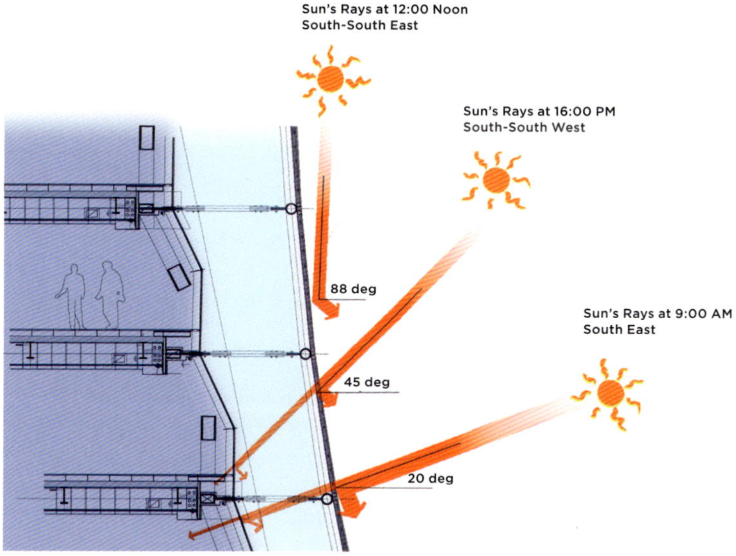

This section through the "splash" sunscreen describes the variances in anticipated reduction in solar gains as a function of the angle of incidence on the stainless steel mesh.

This diagram of the double façade system details the anticipated savings based on the reduction of solar gain and ventilation flow.

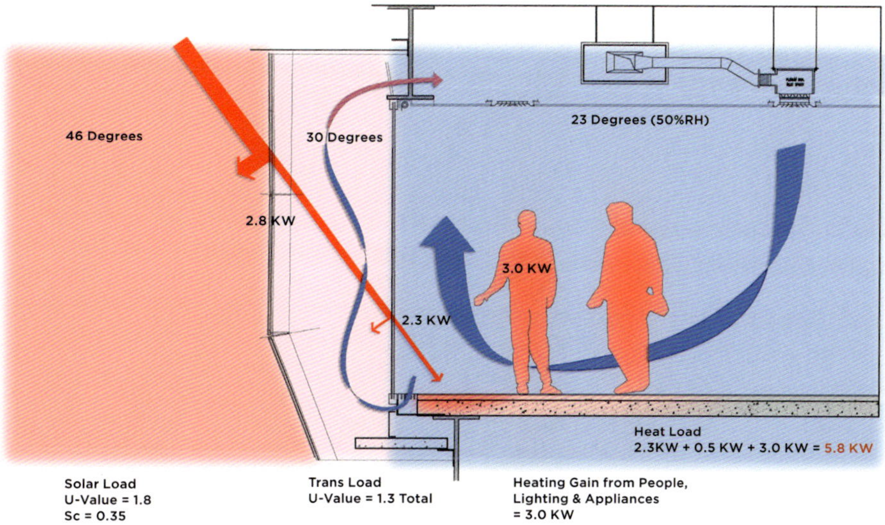

11

PROJECT PROFILES

THE LEADENHALL BUILDING, LONDON, ENGLAND

CAPITAL GATE, ABU DHABI, UAE

GUANGZHOU INTERNATIONAL FINANCE CENTER, GUANGZHOU, CHINA

DOHA TOWER, DOHA, QATAR

ZHONGGUO ZUN, BEIJING, CHINA

LOTTE SUPER TOWER, SEOUL, SOUTH KOREA

The Leadenhall Building in London, England, designed by Rogers Stirk Harbour + Partners with Arup, is the newest diagrid application to be completed at the time of writing. The structure topped out in June 2013. The project employs a double façade system. Here on the south face the grates are being installed between the diagrid members. These will provide support for cleaning access as well as shade.

Where *Chapter 3: Principles of the Contemporary Diagrid* examined how early projects such as London City Hall, Swiss Re and the Hearst Magazine Tower effectively established the precedents that have been used by subsequent buildings in the development of a relatively systematized approach to diagrid structures, this chapter will explore some recent notable examples.

The Leadenhall Building, Capital Gate, Guangzhou IFC and the Doha Tower address a provocative array of design and construction issues. In addition to the comparative, topic-focused information given in the previous chapters, this section is intended to allow the reader to better appreciate the comprehensive approaches to the design of these buildings on the basis of a detailed discussion.

These current Project Profiles serve to extend the information previously provided, in *Understanding Steel Design*, on the detailed design and construction of the addition to the Royal Ontario Museum Addition by Daniel Libeskind and the Bow Encana Tower by Foster + Partners with Zeidler Partnership, which formed the basis for the initial exploration of diagrid constructions.

This chapter also includes more comprehensive information about two visionary projects that remain unbuilt. To date the tallest constructed diagrid tower is the Guangzhou IFC at 439m/1,439ft and 103 floors. The proposals for the Lotte Super Tower in Seoul by SOM and Zhongguo Zun in Beijing by TFP Architects were seeking to exceed this record. Zhongguo Zun is currently under construction, but the competition-winning diagrid design has been abandoned in favor of a megacolumn and concrete core tower under the direction of KPF Architects and Arup. Although the SOM version of the Lotte Super Tower will not be built, the detailed technical development of its design has made a significant contribution to the knowledge base for the application of the diagrid to very tall buildings and will serve to inform diagrid projects that follow.

These profiles were made possible by the generous donation of information and images by the respective architectural and engineering practices involved.

THE LEADENHALL BUILDING, LONDON, ENGLAND

The Leadenhall Building design combines ideas about the relationship of contextual requirements to a structural solution with Architecturally Exposed Structural Steel detailing of the highest quality. The distinctive triangular form that goes under the nickname of "The Cheesegrater" was developed in response to view corridor requirements for St. Paul's Cathedral nearby. The 50-storey building is planned to be 225m/735ft tall. The application of the diagrid structural system is unique and coherent to the extreme that there is no central core. Instead the core has been placed behind the building. The support structure for the tower, predominantly located on the exterior, uses a combination of a diagrid system on its sloped street face with a modified diagonally braced structure on the sides of the building. Arup refers to the total system as a "megaframe", a type that is characterized by the use of large, widely spaced perimeter columns; this type is combined with the diagrid to take advantage of the best qualities of each system. 85% of the building's construction value consists of prefabricated and off-site construction elements, which benefits overall constructability as well as acknowledges issues relating to difficult site access and practical limits on staging area in this dense urban location.

ARCHITECT
ROGERS STIRK HARBOUR + PARTNERS
STRUCTURAL ENGINEER
ARUP
WIND ENGINEERING
RWDI
CONTRACTOR
LAING O'ROURKE
STEEL CONTRACTOR
WATSON STEEL STRUCTURES
PROJECT COMPLETION
TOPPED OUT 2013

The Leadenhall Building makes a striking addition to the historic fabric of the City of London.

The office floors, rectangular in plan, measure 48m/157.5ft in width and up to 43m/141.1ft in depth at the lowest full office floor (level 5). This large floor plate has required the use of interior columns to support the spans. At the same time, the grid is quite large, measuring 16m/52.5ft x 10.5m/34.5ft in order to minimize the number of columns, thereby achieving the degree of openness desired for maximum flexibility in space planning.

The triangular profile of the east and west sides of the tower is subdivided into seven-storey modules. While normally a diagrid module is defined by the measurement from tip to tip of the major diamond shape, which measures 14 storeys on the south façade, the seven-storey designation has more significance to the design of the overall structure because the southern diagrid face is integrated into a megaframe system. Seven 4m/13.1ft-high floors fit within the 28m/91.9ft height that describes the seven-storey module, with each floor 750mm/2.5ft narrower than the previous. The typical floor structure consists of a 150mm/6in deep precast concrete slab over 700mm/2.3ft-deep steel beams. A raised floor system is used to handle services. Arup introduced a passive constrained-layer damping system to reduce the bounciness of the floors.

The plan views of levels 5, 22 and 31 show the placement of the core and services outside of the tower and the open floor arrangement for the offices. This has required the use of columns within the floor plate, which is not usual for more standard arrangements that would see the core and services housed within the floor area.

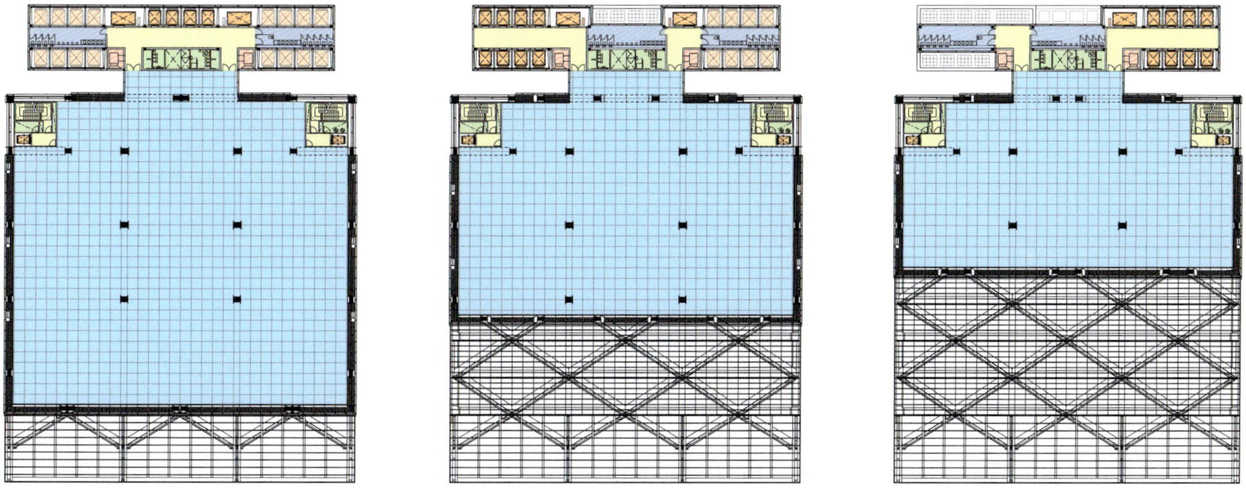

The building consists of seven full seven-storey modules. They sit on a five-storey base that uses a modified structural arrangement to respond to the significantly different program requirements of the galleria space. Here the third and fourth floors are suspended from the fifth floor level and hang within this space. As the perimeter diagonals and columns that comprise the base cannot be braced by the floors, they have been artfully reinforced. The ground level will be accessible to the public.

The primary column elements create a supersized steel frame for the structure that is based on a seven-storey module. The office floors are connected to the seven-storey-long members to provide lateral bracing. There are no additional perimeter columns between these large seven-storey members to provide support to either the floors or curtain wall system.

The ventilated double-skin façade system spans from floor to floor and is supported by the floor beams. It encloses the diagrid/megaframe members via double glazing on the interior and single glazing on the exterior. Placing the structural frame outside of the thermal envelope required that differential temperatures had to be accommodated. Connections of the floor beams to the structural frame via sliding bridge bearings permit small horizontal thermal movements of the frame without transferring these to the floors (see p. 141). Computer-controlled blinds in the cavity automatically adjust in response to solar gain and glare to prevent the glass from heating up. The resulting warmed air is discharged through vents at node level. More information on this double façade system can be found on page 141. Thermal bridging was not as much of a concern in the core area that houses only service functions.

Although the primary structural system would appear to be constructed of wide-flange or Universal sections, all of the members have been customized from welded plate. This allowed Arup to tailor the shapes of the sections to meet the specific requirements of the connections. This construction method results in a more precise appearance of the AESS members. While the detailing of the diagrid and vertical members has been standardized throughout the structure, there is significant variation in the design of the nodes in order to accommodate the different geometries on the three façade types. The nodes measure approximately 6m/19.7ft x 3m/9.8ft and weigh up to 30 tonnes. The splices to the members are moved away from the major load transfer within the node to locate them where the stresses are lower, the geometry is simpler and there is better access and space to accommodate the long boxes that house the tension bolts.

A digital model of the east/west node type illustrating the boxes that facilitate the creation of the prestressed bolted connections. The built version has followed the modeling very precisely, demonstrating the effectiveness of the workflow through the use of digital modeling in the design and fabrication of complex steel structures.

The node connections use high-strength threaded rods instead of standard bolts. Hydraulic force was used to induce tension in these rods in order to make sure that there was enough pressure on the connection. The diameter of the rods ranges up to 76mm/3in and the pretensioning reached up to 200 tonnes.

The bolts transfer their prestress to the ends of the incoming members and the nodes via "bolt boxes", stiffened plates welded between the flanges of the seven-storey megaframe sections. Other plates were attached to the sides of the bolt boxes to transfer the prestress load from the internal stiffeners out to the flanges of the section. These additionally neaten the appearance of the connection. Covers were placed on the ends of the exposed threaded rods to bring the connection up to AESS standards.

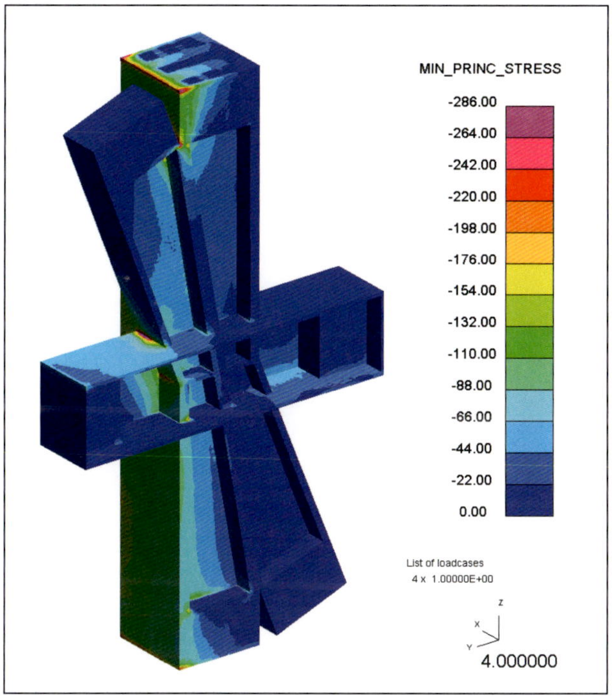

A finite analysis model of a node on the east/west façades allows the validation of the load transfer paths and assists in making decisions on the size and placement of the steel. The nodes must transfer loads of up to 6,000 tonnes in at least three different directions simultaneously.

The megaframe of The Leadenhall Building requires three primary types of nodes to create the connections between the diagonal members. More geometrically complex nodes result where the faces of the frame intersect with each other and with the rear bracing system.

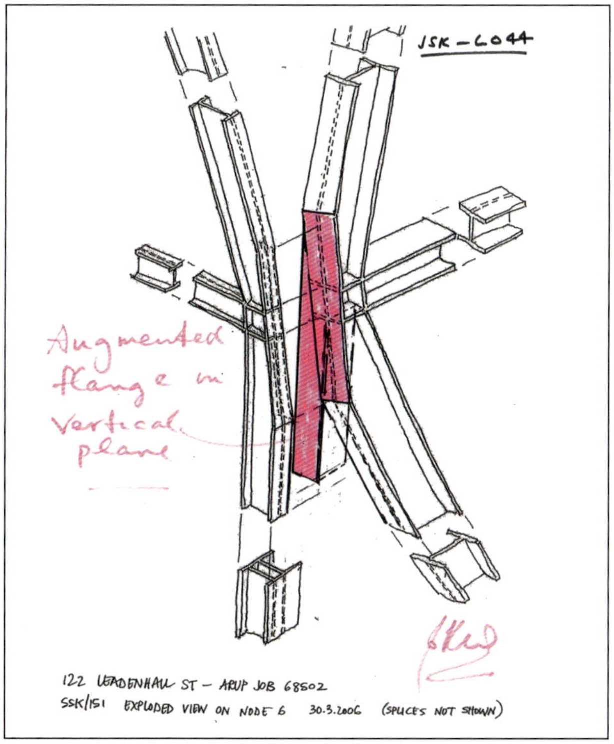

The nodes that provide the transition between the sloped south façade and the vertical sides are highly complex in their geometry. This design sketch shows how the internal plates of the custom-welded members vary in their alignment to accommodate the precise load transfer requirements. Some of the members have two web plates to accommodate the load transfer and create an alignment with the web of the member to which they must attach. Their appearance is consistent with the members that employ a traditional single web.

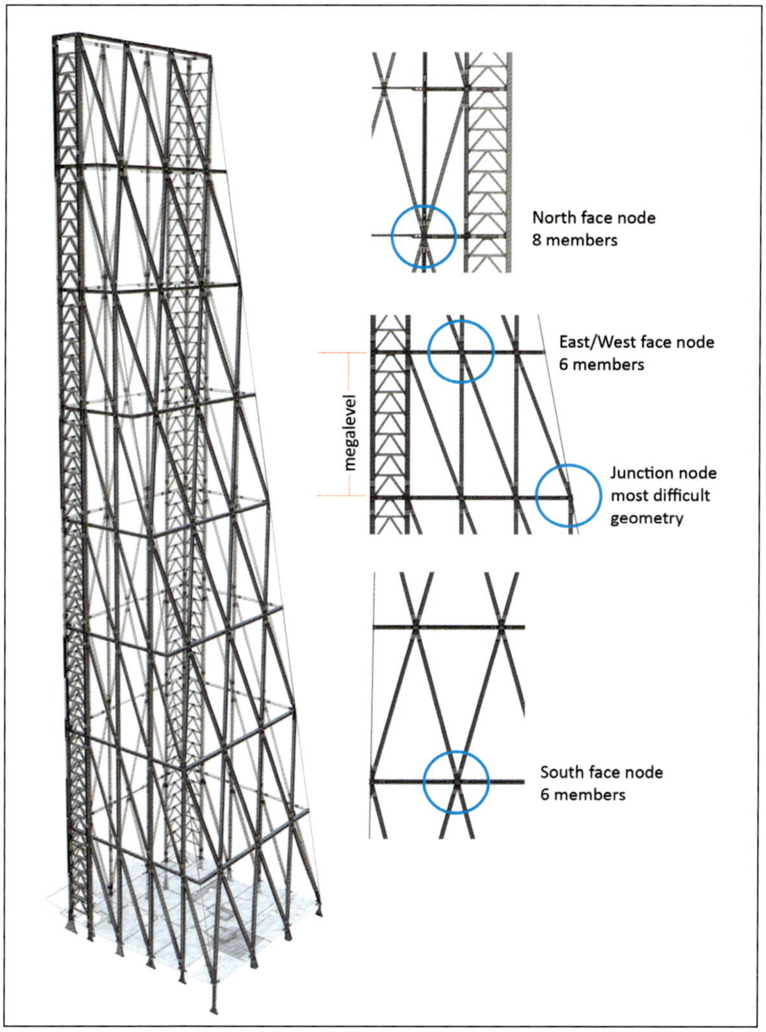

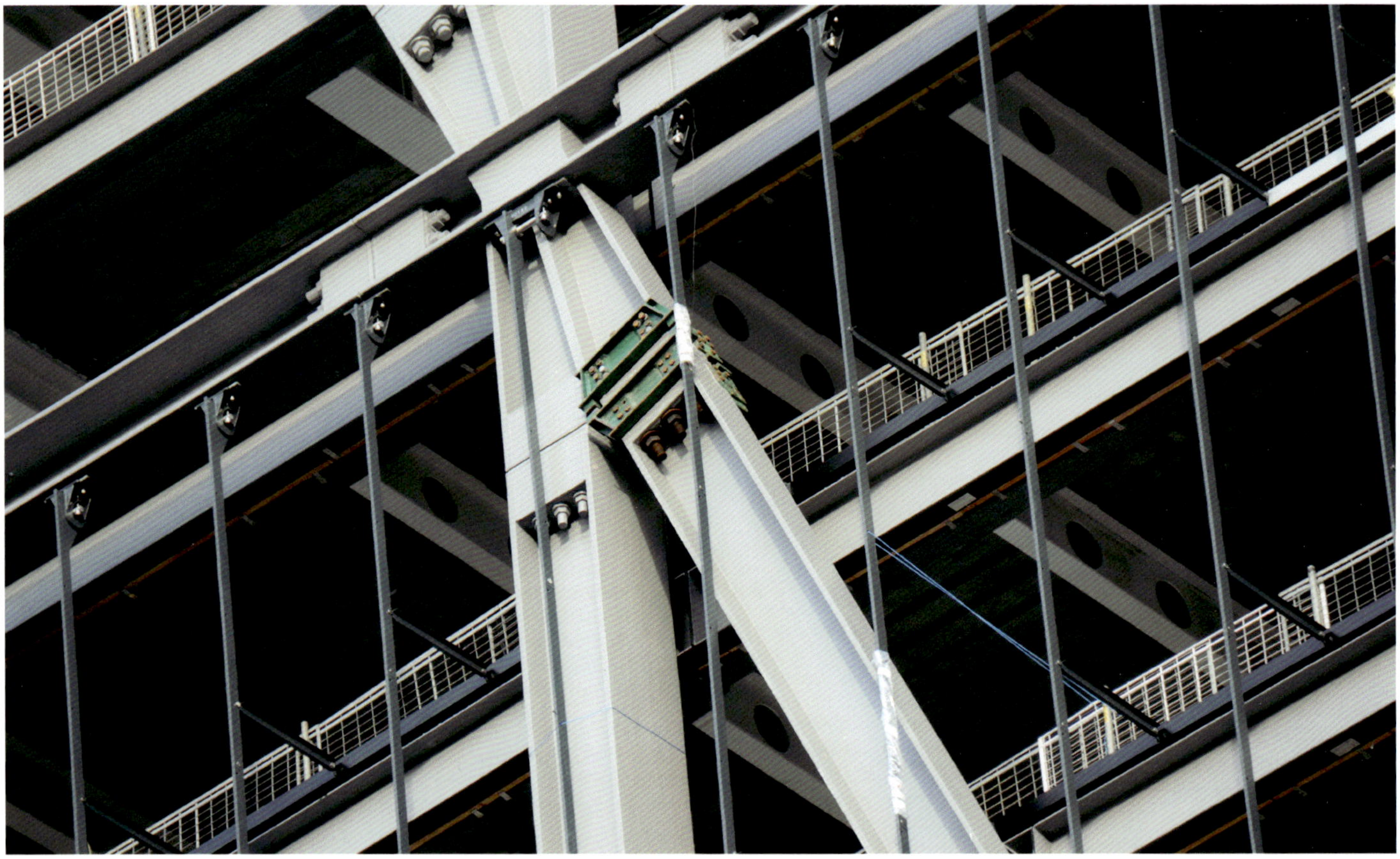

The node type used on the south face is typical of a diagrid node in that it must resolve four diagonals and provide attachments for the horizontal members that define the boundaries of the seven-storey modules. The node type used on the north façade, opposite to the external core, has to accommodate eight members and was more challenging to fabricate due to its unconventional section shape. The nodes at the corners between the sloping south face and the trapezoidal east/west flank faces, on alternate seven-storey "megalevels", have to resolve the most complex geometry. In all there are 12 types of nodes that are derived from a consistent language, based on the accommodation of load transfer and exposed connections.

The nodes on the vertical east and west façades were required to allow modification during the construction of the tower. The design of the connections of these nodes to the frame needed to accommodate load-induced column shortening during construction. The vertical columns on the sides of the building take all of the gravity loads and therefore compress in height as the loading incrementally builds with the addition of floors. By contrast, the diagrid on the south face has no vertical members, and as the geometry neatly wraps around the sloped corners to provide a seamless transition between the two conditions, the verticals are pushed out of alignment with the diagonals, which stay at the same length as installed. This effect was found to push the building over to the northern back side by about 160mm/6.3in. To straighten the tower, the diagonals were periodically shortened to bring the building back into vertical alignment, through a procedure called "active alignment". This was done by loosening the long tensioned bolts in the connection of the diagonals to the node and removing some shim plates. This process also required adaptation in itself to respond to the changing loads that resulted from the diminished floor plate sizes as the tower progressed upward.

A view of a node on the west façade while the exterior frame for the double façade system is being installed. The diagonal is fitted with temporary attachments (in green) that are used together with external jacks to force open the joint in order to take out shims during "active alignment". The closing-up of the joints after shim removal is done by the main jacks tightening the bolts. Most of the joints were in tension during construction, due to the imposed backward lean that resulted from the gravity-induced column shortening on the east/west façades, and therefore in most cases these temporary works were not required. As this is an AESS system, the bolted connections have been designed to allow their exposure.
The white 700mm/2.3ft-deep castellated floor beams that span back to the interior columns are also visible.

The chevron bracing located in the northernmost bay of the tower is designed to stiffen the building between the node levels. The architectural language for the chevron bracing system complements the detailing language of the primary diagrid members. The main connections for the ends of the chevron elements use pin joints. The chevron braces had to be precisely sized as the pin connections must be exact. Discreet bolted connections that sit in recessed pockets allow for assembly of the chevrons on site. The visual lines are kept clean through this detail to achieve an AESS level of finish.

Due to the complexity of the structure, steel contractor William Hare was brought into the process during the design of the structural system under a pre-construction services agreement to advise on constructability issues and to assist in modeling the nodes. Each node type was developed using a simple analysis model to begin to optimize the flow of forces, then modeled using a 3D Tekla model. Although the contract was ultimately awarded to Watson Steel Structures, the early modeling work provided continuity for the fabrication processes. Watson Steel was able to create highly accurately machined connections for better bearing contact, thereby reducing welding requirements in the nodes and simplifying the shimming strategy.

The south façade of the building employs a classic diagrid frame. There are no vertical columns. The long diagonals are braced at the respective node by substantial beams that also serve as the termination points for the double façade system that spans each seven-storey megalevel of the building. Deep castellated floor beams span back to the columns in the interior. The beams form the edge of the interstitial floors and attach to the diagonals to provide lateral bracing by reducing their unsupported length. The geometry wraps around the corners, where the system shifts to a column-and-diagonal-brace system.

The large scale of the modules and the placement of the core to the north of the tower necessitated the introduction of an additional stability system to stiffen the structure between the node levels. A chevron or "K-bracing" system was introduced in the northernmost bays on the east and west faces. These enclose two sides of the fire-fighting cores, which were required to be situated within the primary floor area. The chevron system was chosen because it can accommodate the shortening of the columns on the sides and back during construction without attracting large forces into the braces.

The service core has been fabricated from a highly engineered set of prefabricated components. These can be very light in their appearance and connection types as the core is not required to assist with lateral stability.

The chevron bracing system wraps around the corners of the tower. The construction of the core was able to proceed relatively quickly due to the extensive use of off-site prefabrication for its components.

Fire protection on the architecturally exposed steel structure was achieved through a shop-applied marine-standard epoxy-intumescent coating on the exposed steel frame. The thickness varies from 3mm/0.125in to 12mm/0.5in as a function of the thickness of the steel – thinner steel requires thicker protection to achieve the same rating. The rating achieved will be 90 minutes. Epoxy-intumescent coatings are normally shop-applied and this necessarily infers more careful handling during transportation and erection. Still, some damage to the finish is unavoidable, so a permanent artist touch-up person was on site to remediate damage as it arose.

Many thanks to Damien Eley, a structural engineer with Arup, for his assistance in providing technical information for this project.

The Leadenhall Building sits above the adjacent towers in the City of London (right), the grouping creating a new definition of tall building for the city.

The curtain wall installation (left) underway on the north façade of the service core. The prefabrication of components was highly beneficial to this building element.

CAPITAL GATE, ABU DHABI, UAE

Capital Gate made the Guinness Book of World Records for being the furthest leaning man-made tower for its 18° lean. The design team for the 165m/541.4ft, 35-storey tower undertook an innovative and exploratory approach to tackling a number of issues associated with the iconic design that was desired by the client. In the words of Jeff Schofield, formerly of RMJM and one of the designers of the building: "It was the right time in history and we had the right technology to make this project happen." The use of a steel diagrid was critical to the success of the structural stability of the building as well as to its visual appearance. 18 diagrid triangles encircle each level of the building. With its shell-like behavior, the diagrid provides a perimeter structure that is adept at conforming to the varying vertical geometry. This has eliminated the need for columns that would otherwise have had to change positions and intrude in the interior spaces. Conventional framing would not have been suitable.

ARCHITECTS
RMJM

STRUCTURAL ENGINEERS
RMJM

PROJECT DEVELOPER
ADNEC (ABU DHABI NATIONAL EXHIBITIONS COMPANY)

PROJECT MANAGERS
MACE

MAIN CONTRACTOR
AL HABTOOR ENGINEERING ENTERPRISES

STEEL CONTRACTOR
EVERSENDAI

FAÇADE CONSULTANT
HYDER CONSULTING

FAÇADE CONTRACTOR
WAAGNER BIRO

COMPLETED
2011

Capital Gate has used the strength of the diagrid to assist in achieving the iconic nature of this project.

The complexity of every aspect of the design, fabrication and erection of this project would not have been possible without the use of digital modeling. The ability to translate the complex geometries established in the design model through to the engineering and fabrication models was explored at a very high level in this project. Files exported from Tekla Structures facilitated faster and more accurate fabrication. For the fabrication of diagrid members, temporary jigs were made with the help of coordinates of individual assemblies that were taken from the model. Fit-up of the individual parts of the nodes was done with the 3D coordinates taken from the model. The location of the center of gravity and weight of each individual assembly were taken from the reports generated by the software. These were used to work out the lifting methods and also in the design of the erection engineering.

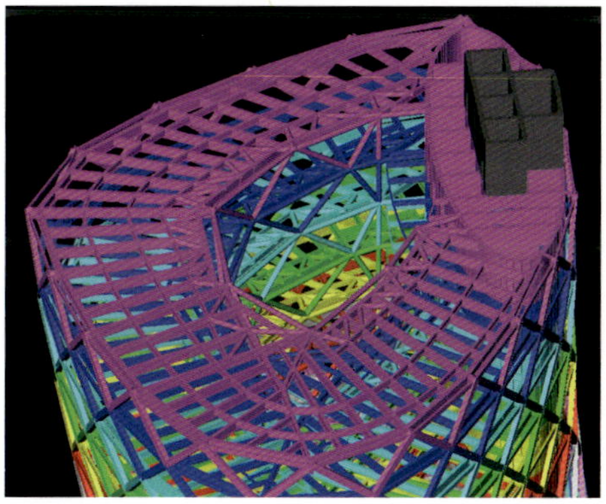

A closer screen view of the Tekla model used for the design of all of the structural steel. This view is taken through the upper atrium and shows the offset of the hotel floor area from the core.
The digital models use variations in colors to assist with reading the model as well as in some cases to denote erection sequences.

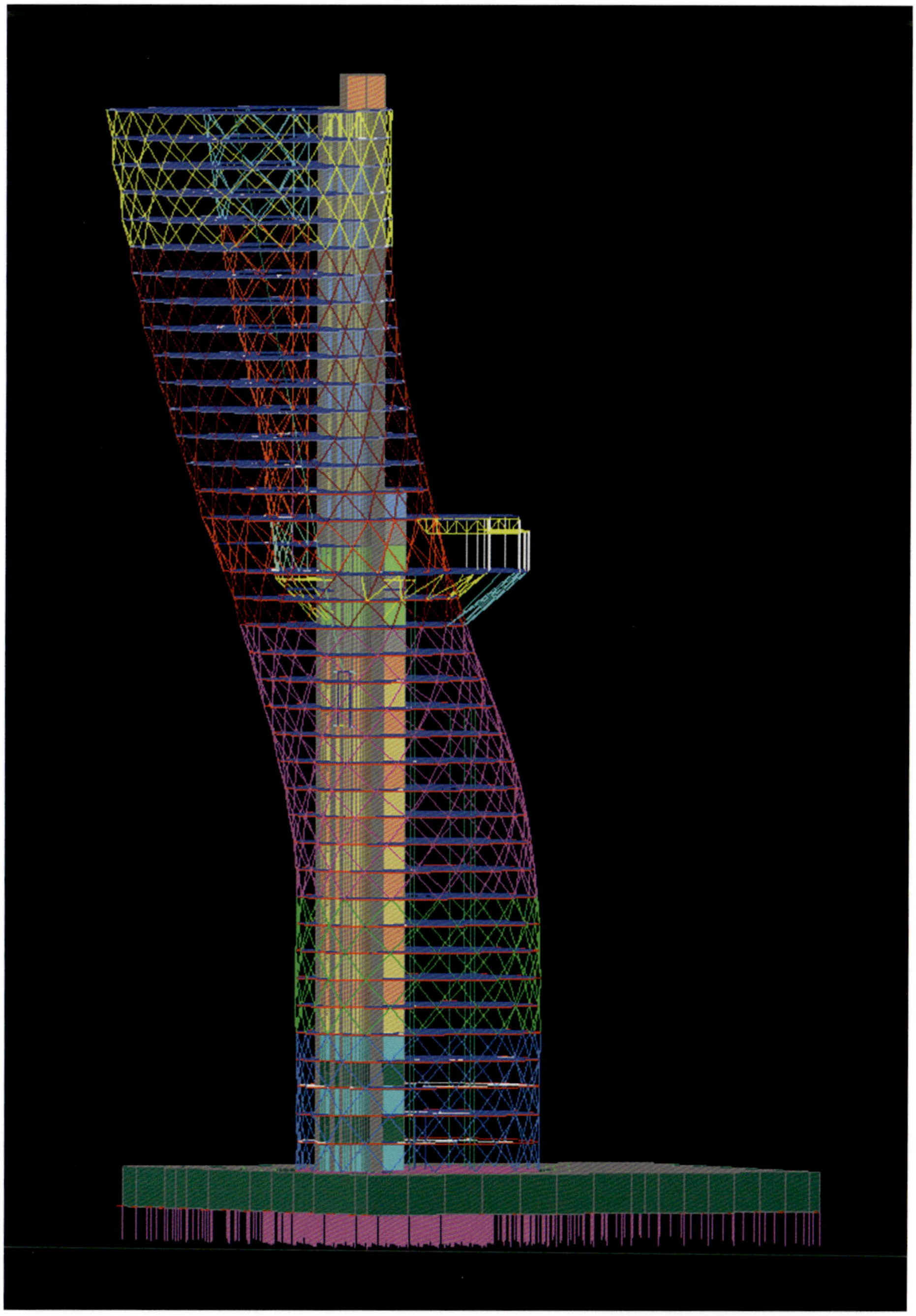

The Tekla Structures steel modeling system was used extensively on this project. This is a screenshot of the entire model which shows the positioning of the core with respect to the diagrid.

This digital rendering illustrates the positioning of the concrete core in relation to the curvilinear shape of the exterior diagrid. Steel outriggers at the mid-height (pool) level assist in countering the eccentric weight of the hanging hotel floors at the upper level.

This detail of one of the base nodes indicates the method of attaching the node to the diagrid member. Temporary plates are used to secure the members with bolts. These must be removed as well as the weld marks, if the structure is to be exposed as AESS. If the splice welds remain visible, they will be ground smooth. The diagrid structure is ultimately coated with an intumescent material for fire protection, which can also help to mask some imperfections.

Custom-fabricated metal covers are used to hide the bolted connection between the horizontal tension hoops and the nodes. The covers will be filled and sanded and when finished will blend in with the appearance of the node. Not all of the diagrid members are fabricated purely "square". As can be seen here, some of the diagonals that must resist high loads have been reinforced with additional material. The use of plates to connect the incoming members assists greatly in solving the varying geometries of the connections as well as making an attractive detail.

The diagrid elements were fabricated by Eversendai at plants in Sharjah and Dubai with steel mainly from Europe. The actual cut-to-length measures of the members were taken from the preliminary model developed at the initial stage of the project to procure the material. This helped in reducing material wastage. To maintain the required curved surface profile, every panel surface is slightly deviating with respect to the adjacent panel surfaces. External diagrid members are fabricated as hollow sections that measure 600mm x 600mm/2ft x 2ft. The steel thickness of the diagrids varies from 80mm/3.2in at the base of the tower to 40mm/1.6in at the top where the loads are lighter. Custom-welded sections are used at the base of the building where the loads are high and thickness requirements exceed standard material. Standard HSS tubes are used toward the top where the loads are lighter and the section requirements can be met more economically with standard material.

The diagrid that frames the double-height lobby at the ground floor clearly expresses the differentiation in the member shape of the horizontal hoop, or tension tie, that is used to triangulate the diagonal grid.

The most severely leaning diagrid is located on the 18th floor at the restaurant level. The diagonal members are visibly reinforced. The façade cladding system has already been installed and work is proceeding to clean up the welded connections.

Wide flange or Universal members frame into the beams that tie the diagrid to the core. This system will be fire-protected using spray fireproofing as it will be concealed.

Custom-fabricated beams are used to tie the diagrid back to the concrete core. The ends of the beams employ a pin connection.

The diagrid members of the interior atrium are made of circular hollow tubes and fabricated as large X-shaped elements that were welded together on site, with the horizontal welded connections exposed to view. This required a high level of precision in the transfer of information from the digital model to the fabricated steel, as the tolerances for alignment were very tight. The structure as formed by the interior and exterior diagrids, connected by the steel floor beams, needed to be very resistant as it was separated from the core by the void created by the atrium.

The hotel atrium uses a different structural language for its diagrid. Round HSS material is joined using cruciform steel plates to create the geometric transition. This detail adds interest as well as alleviates the difficulty of joining four tubes of varying angles at one connection. The grey zones visible on the tubes denote the splice zone for the X-shaped diagrid elements. Using a system of X-shaped elements comprised of nodes and diagonals significantly reduced the number of on-site welds.

Large steel truss outriggers are used to support the pool level at the 19th floor that serves to terminate the "splash" sunshading device on the entry side of the tower.

The design of the core required innovation in order to accommodate the extremely eccentric loading. During construction the core was "pre-cambered", meaning that it was poured approximately 35cm/1.12ft off vertical so that when the floors were framed in on the opposite side, the additional load would straighten the core. The core was also post-tensioned with vertical cables on one side to counteract the lean on the other side. The cables were installed in vertical segments that overlap each other every seven floors.

On a highly complex project, construction sequencing is carefully staged to allow the construction to progress at a reasonable pace. Here, the cladding installation is several levels behind the finish welding on the diagrid, which in turn is several floors behind the pouring of the concrete core.

Each node for the building was unique as it had to accommodate ever-changing geometries. Here a node is still attached to the crane after a lift. Such extensive welded connections required the construction of scaffolding at each connection to provide a safe condition for working at height.

Capital Gate succeeded in stretching the limits of diagrid construction through its unusual geometry. The project would not have been possible without extremely advanced digital design tools and a highly adept team. The project also explored a range of sustainable strategies. For information on the sustainable strategies and double façade construction please see page 144.

Many thanks to Jeff Schofield of ADNEC for assistance with this project.

GUANGZHOU INTERNATIONAL FINANCE CENTER, GUANGZHOU, CHINA

The tower in the Central Business District of Guangzhou was completed in 2010, has 103 floors and reaches a height of 437.5m/1,435ft with a curved triangular plan. As of 2013, it is the tallest building incorporating a diagrid structure. The lower 69 floors are flexible office space and the upper 34 storeys host a Four Seasons Hotel. The hotel reception on the 70th floor opens up to a central atrium.

ARCHITECT
WILKINSON EYRE ARCHITECTS

ASSOCIATE ARCHITECTS
ARCHITECTURE DESIGN INSTITUTE OF SOUTH CHINA UNIVERSITY OF TECHNOLOGY

STRUCTURAL ENGINEERS
ARUP, ARCHITECTURE DESIGN INSTITUTE OF SOUTH CHINA UNIVERSITY OF TECHNOLOGY

PROJECT MANAGER
YUEXIU GROUP (FOR FOUR SEASONS HOTEL)

MAIN CONTRACTORS
CHINA STATE CONSTRUCTION, GUANGZHOU MUNICIPAL CONSTRUCTION GROUP, JOINT VENTURE

COMPLETED
2010

"The design for the Guangzhou International Finance Center makes the case for an aesthetic that aims for an elegant simplicity, but expresses the function, structure and components of construction in a clear and direct way." (Chris Wilkinson)

The design, a collaborative effort between Wilkinson Eyre Architects and Arup, breaks fresh ground to create a light structure. Developing this Supertall building in the typhoon climate of China's south coast posed great challenges to Arup's design engineers. Extensive computer analysis was conducted to find the optimum geometry for the diagonals and the floor layout in relation to the curve of the building elevation and profile. The designers sought to create a visual contrast between the exposed diagrid structure and the smooth curved form of the all-glass cladding.

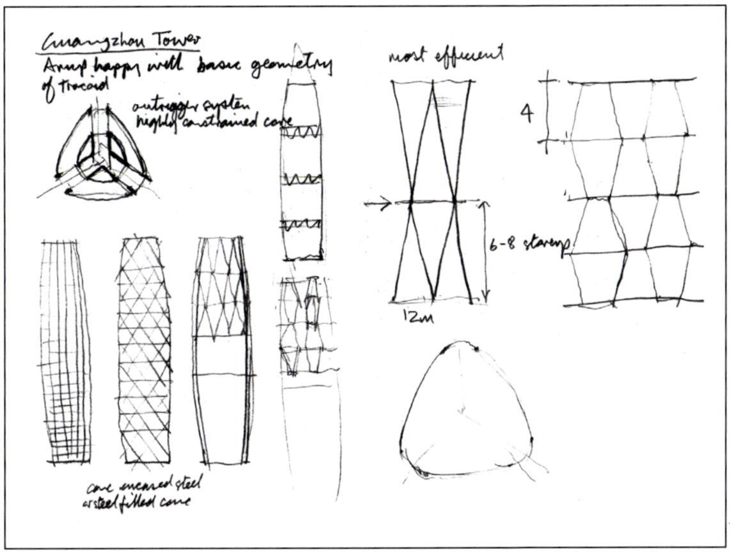

Developmental sketches describing the design process that created the shape of the tower and the selection of the modular diagrid structural system.

In section, the external walls curve out from the ground floor to a maximum girth at approximately one third of the height, then taper gently to the top. According to Wilkinson, "the resultant form looks deceptively simple but is based on a complex toroidal geometry, with each of the three façades set out on a radius of 5.1 km/3.17mi vertically and with a radius of 71m/233ft in plan and 10m/32.8ft on the corners." The refinements to the design were determined by the use of rapid prototyping with physical models as well as computer modeling.

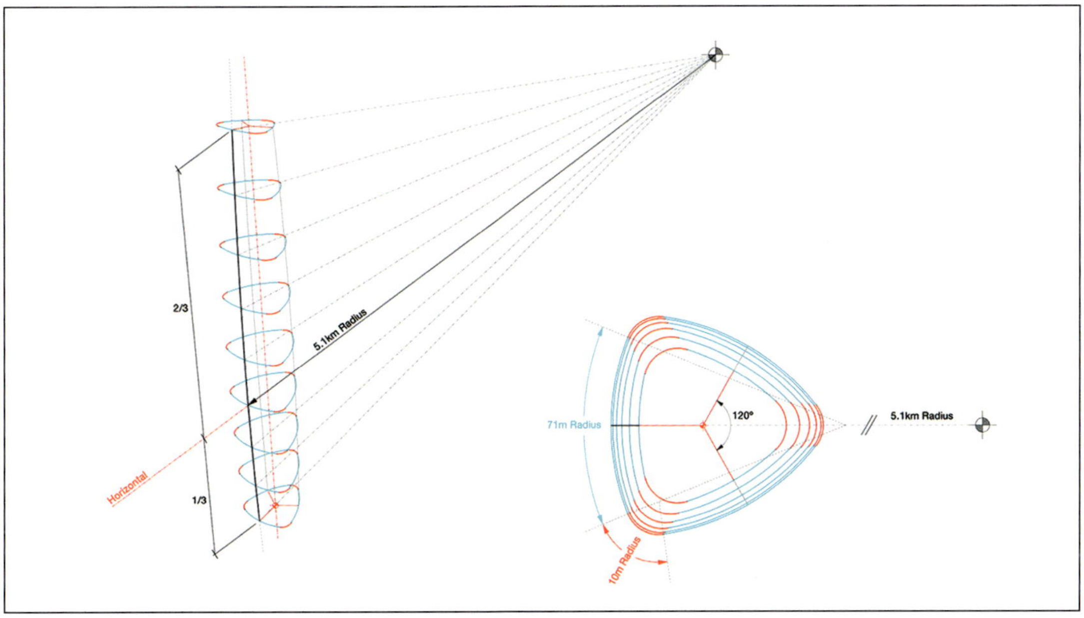

The tower is located in a monsoon area, so wind issues dominated many of the decisions. The plan, which is a curved toroidal triangular shape, was developed to deal efficiently with external wind forces. The gracefulness of the curves is the result of the use of large radii.

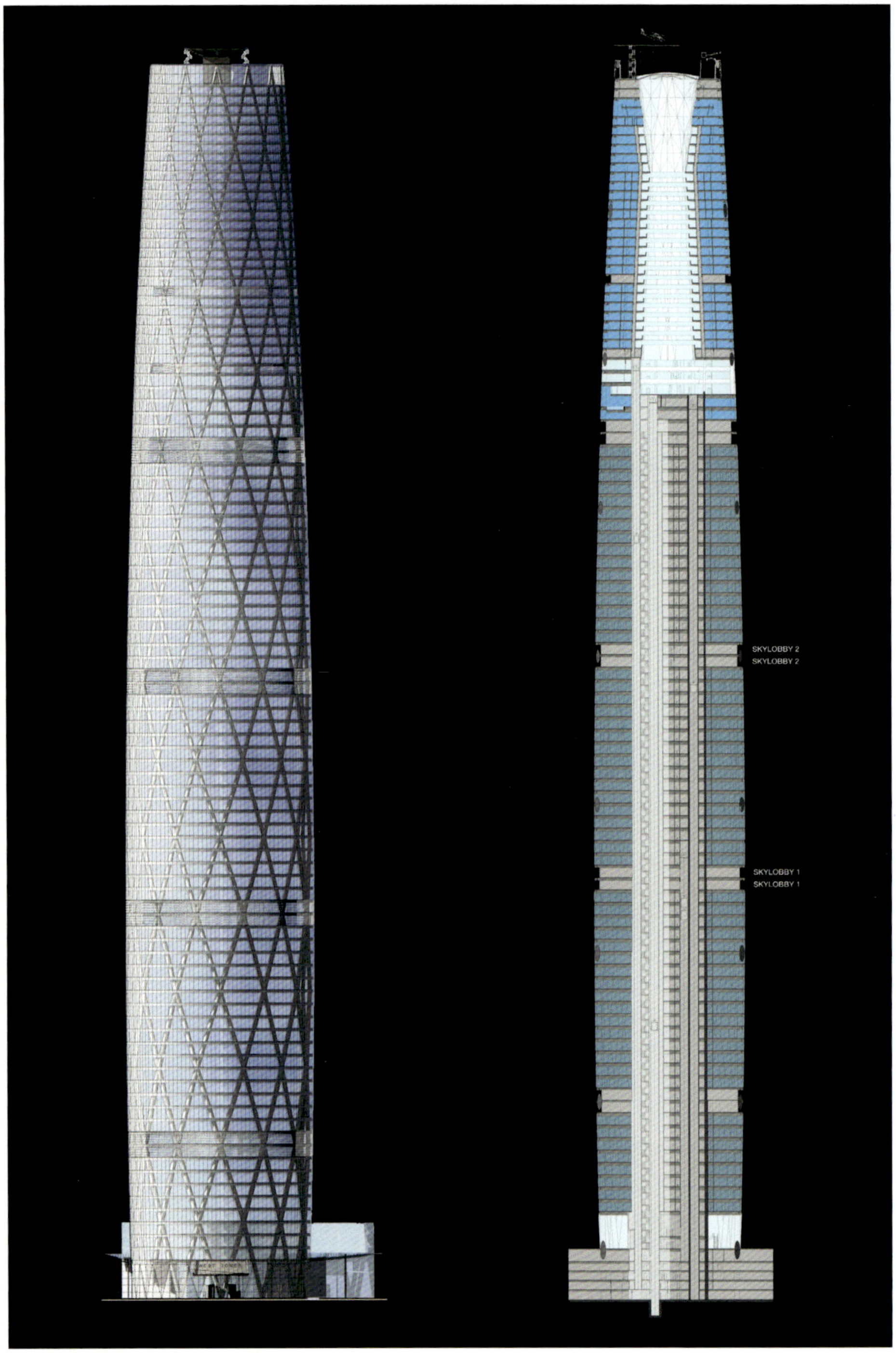

The elegant simplicity of the diagrid framing system is clear in this digital model. The floors provide bracing for the long diagonals along their length, instead of relying only on a hoop system that would primarily connect the nodes. This is required as a result of the large module size.

The elevation reveals the delicate geometry of the curve as the tower swells slightly at the mid-section and narrows toward the top. The section view reveals the remarkable change in the upper hotel floors allowed by the displacement of the concrete service core.

The perimeter diagrid system vastly reduces the amount of steel required, and according to the engineers it is also a good arrangement for seismic design. The diagrid behaves as an external tube that is fully braced, so that most of the forces are transferred as axial forces, which makes it more efficient than a moment frame. Optimizing the geometry for efficiency and visual appearance has resulted in a "giant order" of diamonds, 54m/177ft tall, spanning between 12 office floors and 16 hotel floors (the floor-to-floor height requirement of the hotel being less than the office). The dimensions of the module and the floor-to-floor heights were designed to keep the module height consistent throughout the height of the tower.

Looking up the triangular atrium on the hotel levels toward the skylight. This central daylit space is directly related to the ability to terminate the structural core at the 70th floor/hotel lobby level and move these functions to the corners of the triangular plan.

The diagrid, combined with the central reinforced concrete core, provides both gravity and lateral resistance. There has been no need for dampers to reduce horizontal movement. The central core, which serves the offices (see p. 107 for the central atrium in the upper part of the building), was constructed with a "climbform" system. Its shape was determined by the lift configuration and the briefing requirement to provide flexible access for multiple letting of the office space. There has been no need for dampers to reduce horizontal movement. The flexibility has facilitated the creation of the large central atrium at the hotel levels without requiring any additional major bracing structure.

The atrium at the hotel lobby level on the 70th floor is supported by another diagrid structure to create a consistent design language. The diagrid members are significantly reduced in scale. The diagrid in the foreground supports the inner edges of the floors that front onto the upper atrium.

The concrete-filled steel tubes at ground level are sized to resist the cumulative load of the 103-storey tower. They are simply finished to complement the interior design of the lobby.

DOHA TOWER, DOHA, QATAR

The 238m/781ft tall Doha Tower is a cylindrical volume that measures 45m/148ft in diameter. Standing 46 floors, it is crowned by a dome that ends with a light tower at 231.5m/759.5ft. The concrete-filled steel tube structure follows a diamond-shaped grid that bends along the virtual surface of the cylinder. The perimeter diagrid is offset inside the cylindrical shape and is set back from the curtain wall with distances that vary, being closer on the north side and more distant on the south side. The diagrid members vary in diameter from 1.6m/5.3ft diameter on the 14th floor to 800mm/2.6ft by the 27th floor level to reflect differences in loading.

ARCHITECT
ATELIERS JEAN NOUVEL

STRUCTURAL ENGINEER
TERRELL GROUP; CHINA DESIGN CONSTRUCTION INTERNATIONAL

MAIN CONTRACTOR
CHINA STATE CONSTRUCTION ENGINEERING CORPORATION

COMPLETED
2012

South façade of the Doha Tower showing the highly detailed outer skin of the double façade. The mashrabiya is more dense on this façade to limit solar penetration.

The façade uses a double-skin system (for a detailed description see p. 142). The application of the diagrid system on this building is quite unique in that the mashrabiya shading system of the double façade obscures the impression of the diagrid from view during the daytime, but allows it to be revealed during the nighttime hours, when lit from the lighting system that is supported by the grated floor system within the double façade cavity. The following illustrations by Ateliers Jean Nouvel (photographs © CSCEC) allow a detailed look at the special features of this tower.

The plan of the 14th floor shows the offset of the core to create more significant consolidated space. The 1.6m/5.3ft diameter of the diagrid members reduces to 800mm/2.6ft by the 27th floor level. The diagrid is also eccentric and more widely separated from the façade on the south face of the tower, to allow the diagrid elements to act more sculpturally on the interior.

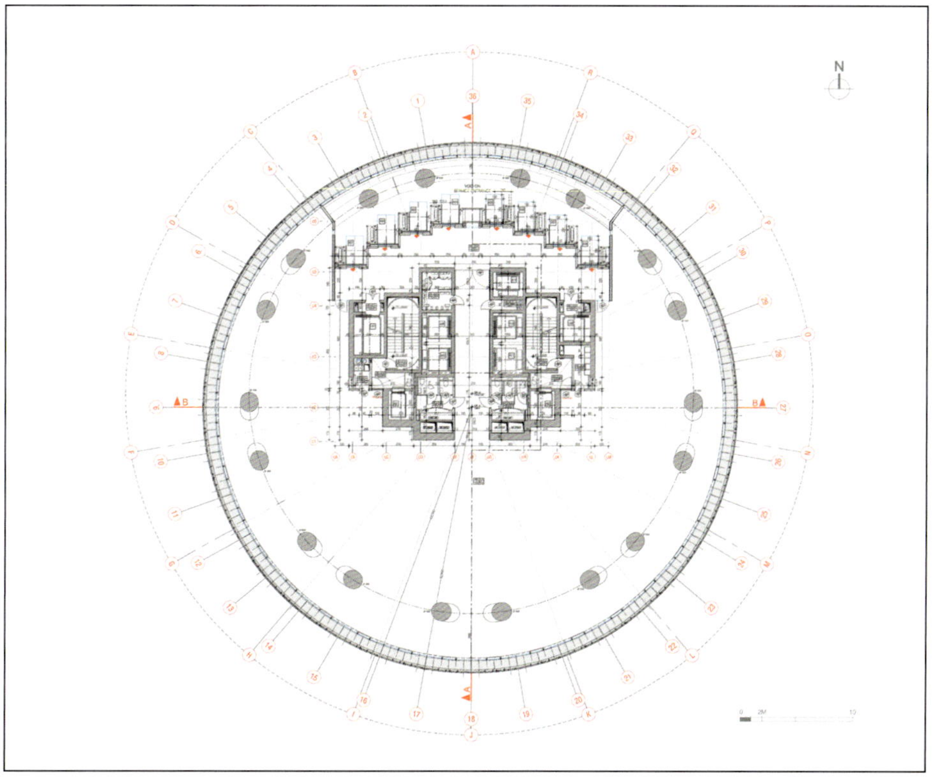

The large diagrid support system is expressed in contrast to the fine texture of the sunlight as it comes through the mashrabiya shading system. The strength of the perimeter diagrid system has allowed the core to be offset, creating very open floor areas. This infers that the diagrid is assuming the majority of the lateral loading.

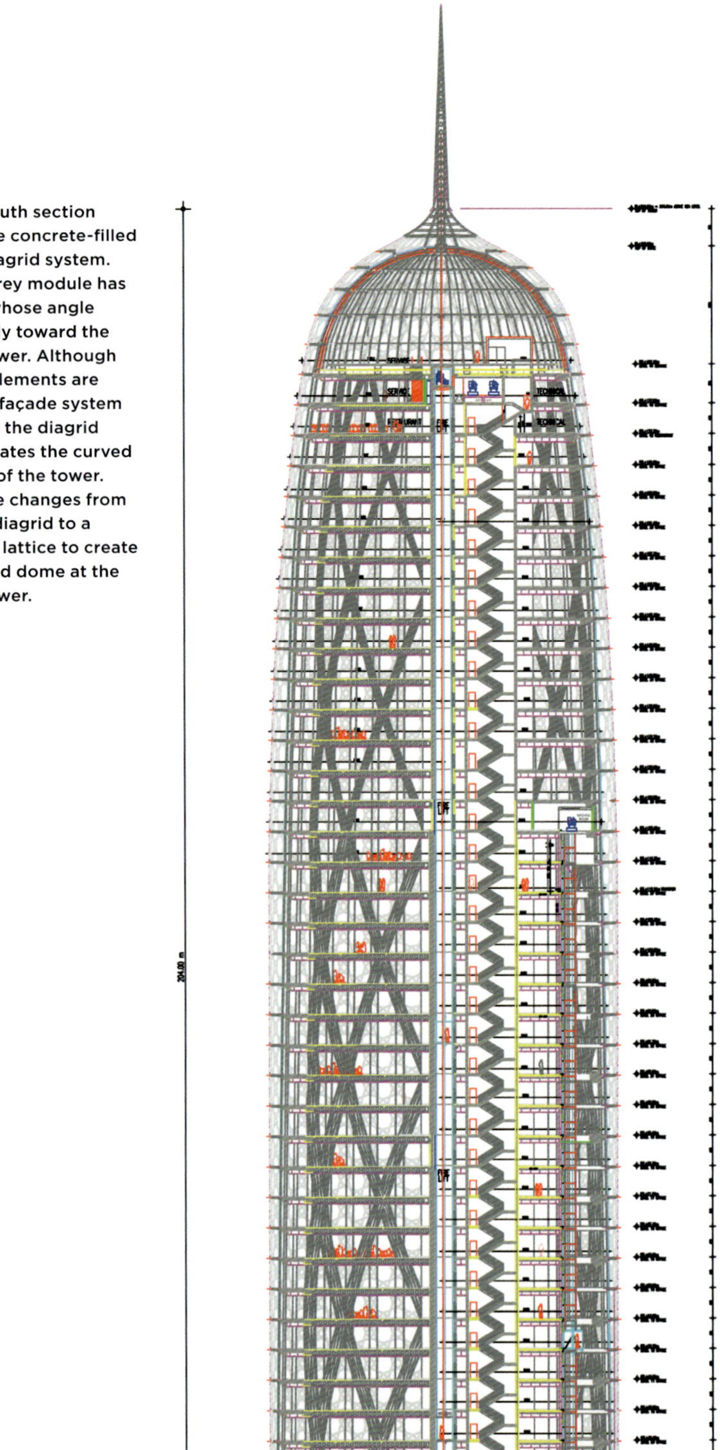
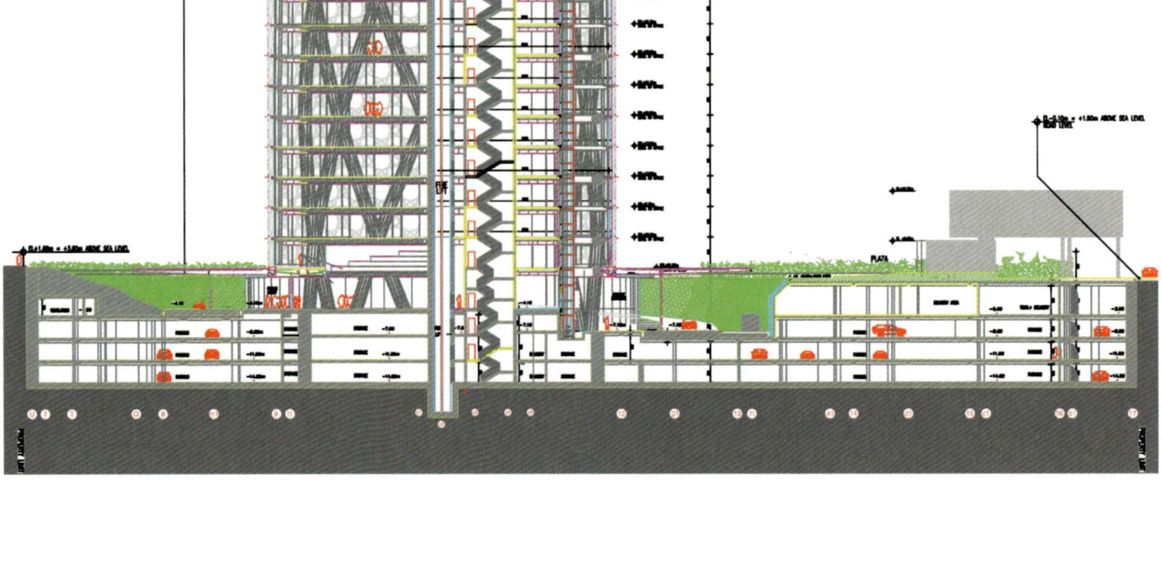

The north-south section illustrates the concrete-filled steel tube diagrid system. An eight-storey module has been used, whose angle tapers slightly toward the top of the tower. Although the diagrid elements are straight, the façade system as applied to the diagrid structure creates the curved appearance of the tower. The structure changes from a structural diagrid to a lighter AESS lattice to create the light-filled dome at the top of the tower.

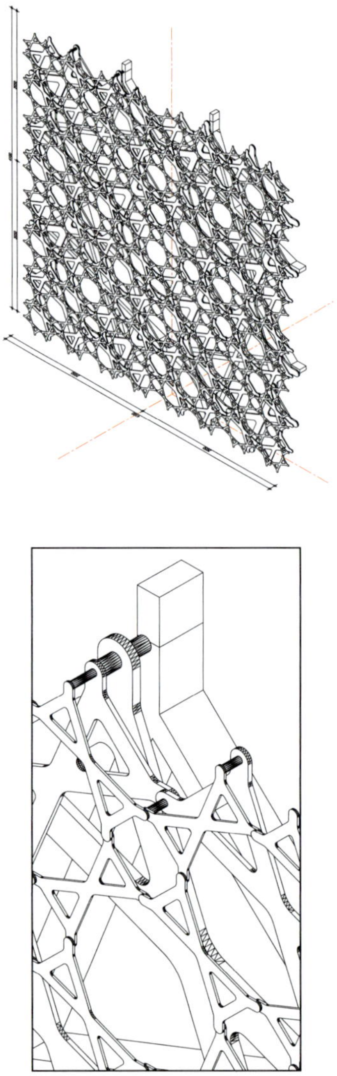

Isometric details of the mashrabiya screen, comprised of several layers of aluminum. The addition of layers results in a denser screen and increased shading.

Details of the screen overlay system (left) used on the exterior of the tower. A cross section (right) indicating the system of support and scale.

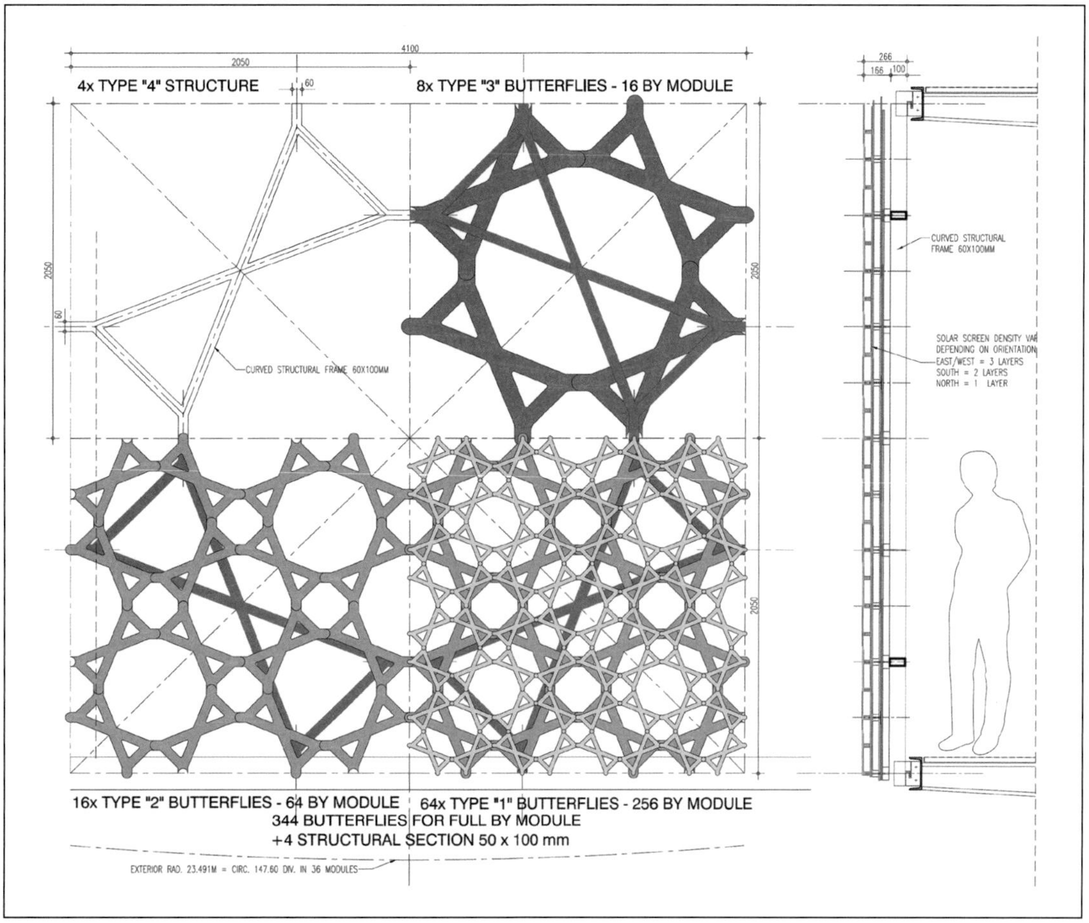

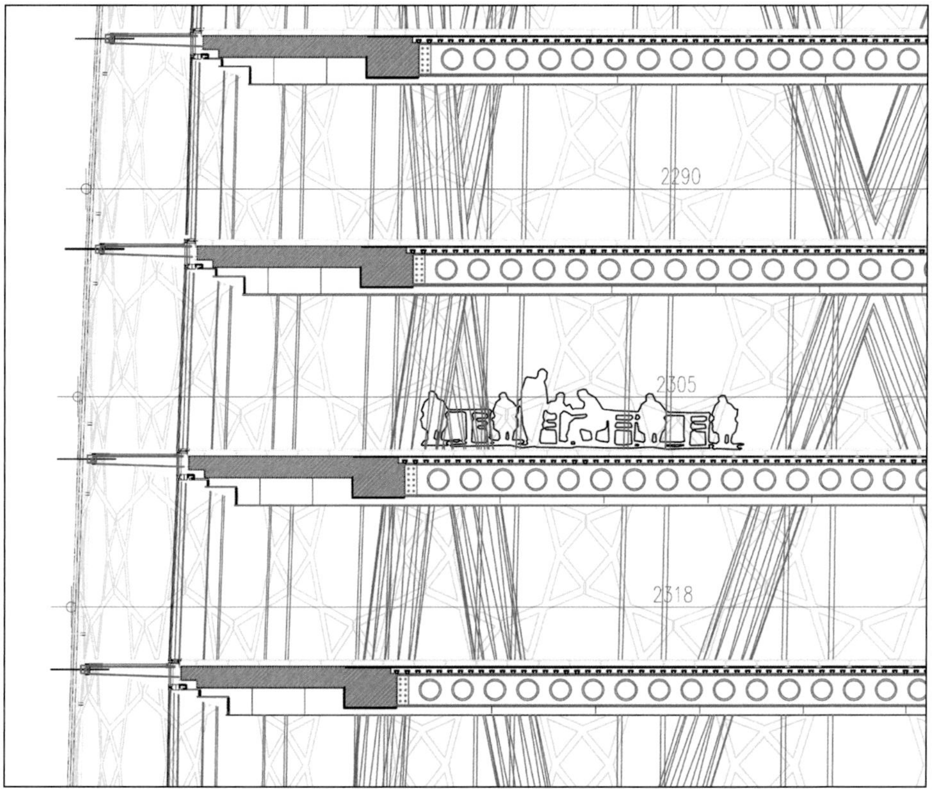

A cross-sectional view through the office floor areas showing the relationship to the exterior façade screen. The diagrid elements are set back from the façade, at a distance that varies around the perimeter as a function of orientation and also throughout the height of the tower. Adequate space has been left between the mashrabiya screen and the curtain wall to allow access for cleaning. A steel grating system at each floor level provides for this access and also houses the lighting system that is used to animate the tower at night.

ZHONGGUO ZUN, BEIJING, CHINA

The competition-winning design by TFP Architects for the Z15 or Zhongguo Zun Tower in Beijing proposed a perimeter diagrid to create the unusual shape of the building. The planned 108-storey, 528m/1,732ft tower will be the tallest in Beijing once complete. The hourglass shape took its inspiration from the ancient Chinese Zun wine vessels. Construction on the project started in 2013, with KPF taking the lead on the project with a modified proposal that did not use a perimeter diagrid support system. Fire engineering was a prime concern, in particular due to the anticipated occupant load at the upper observation level of the building.

The proposal by KPF selected for realization is based on a dual system for lateral force resistance, composed of a fully braced megaframe and a concrete core. Composite steel-concrete material is used extensively to minimize structural member size and increase the usable floor area. The building is designed to set the new height record in a high seismic zone.

COMPETITION ARCHITECT
TFP ARCHITECTS

PROJECT ARCHITECTS
KOHN PEDERSEN FOX ASSOCIATES

STRUCTURAL ENGINEER
ARUP

FIRE ENGINEER
ARUP

PROJECT DEVELOPERS
WHARF HOLDINGS, CITIC PACIFIC GROUP

An aerial rendering of the competition-winning proposal for the Z15 Tower, showing the tower as the focal point for a substantial new development in the Chaoyang District of Beijing.

A view of the base of the tower proposal, showing the design concept of the diagrid modularity in combination with a system of shades that highlight the diagrid pattern.

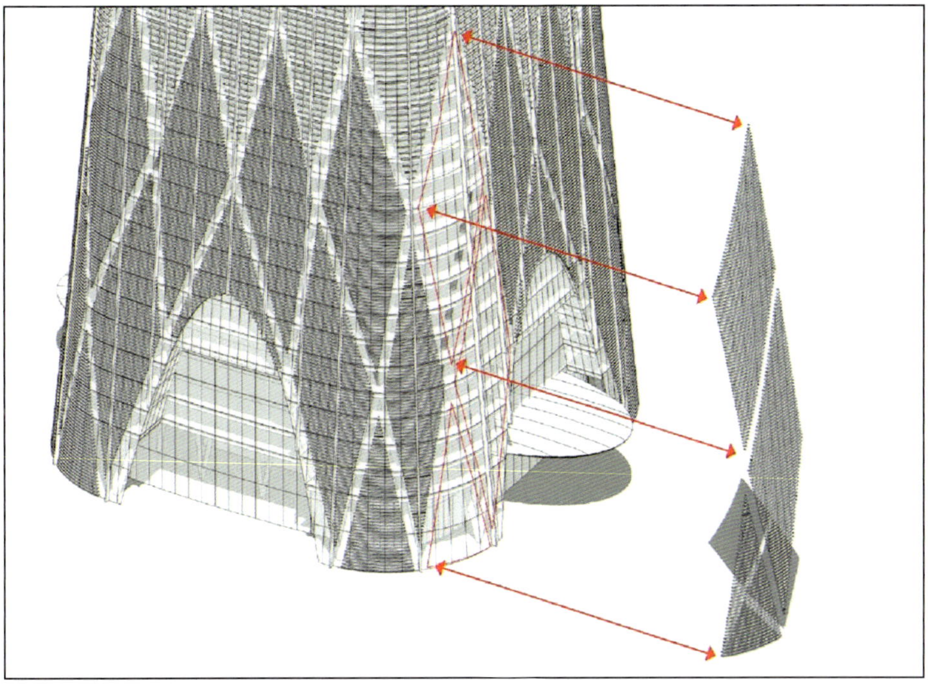

Proposed expression of the diagrid at the top of the tower.

A proposal for the interior of the tower, demonstrating a potential expression of the diagrid structure in one of the lobby areas.

LOTTE SUPER TOWER, SEOUL, SOUTH KOREA

Although this proposal for the 555m/1,821ft-tall Lotte Super Tower is not to be built, it represents an essential step in the detailed development of the diagrid structural system. Unlike many form-driven tower proposals, this team has explored the potential and technical details of the geometries of the diagrid system in great depth. This Supertall building would have become the tallest diagrid tower in the world.

The proposed 112-storey tower employs a unique form that transforms from a 70m/230ft square footprint to a 39m/128ft circle at the top. Geometric morphing from a square base to a circular top posed technical challenges for the structural engineering team but also provided significant benefits. The shape of the building was wind-engineered to reduce wind forces and accelerations. The change in geometry along the height of the tower results in a steeper inclination of the diagonals. According to William Baker of SOM, "the diagrid geometry reflects the 'tall building problem' - flatter diagonals at the top where shear dominates and steeper ones near the base where overturning moment dominates. It was important to the design team to tell the story of a tall building honestly." The rounded shape at the top also responds to wind design issues by removing the corner conditions to reduce the vortex shedding. Vortex shedding was additionally countered by designing the diagrid at the top floors as an open trellis.

ARCHITECT
SOM SKIDMORE, OWINGS & MERRILL

STRUCTURAL ENGINEER
SOM SKIDMORE, OWINGS & MERRILL

PROJECT TEAM
WILLIAM BAKER, CHARLES BESJAK, BRIAN MCELHATTEN, PREETAM BISWAS, BONGHWAN KIM, ARKADIUSZ MAZUREK

WIND ENGINEERING
RWDI

A sectional model of the top of the tower showing the open trellis that is used to reduce vortex shedding. The type of profile used for the horizontal tension rings is clearly differentiated from the cross section of the diagrid members. A characteristic benefit of a diagrid tower is the ability to reduce dependency on the concrete core at the top of the tower where shear loading dominates. This permits the creation of high-quality architectural spaces.

A rendering of the proposed Lotte Super Tower.

A dual system comprised of an exterior steel diagrid and a ductile interior reinforced concrete core was proposed for lateral resistance to wind and seismic effects. The reinforced concrete core increased the mass of the building to limit building wind motion. Comparisons were made between the diagrid solution and a more conventional perimeter moment frame solution. It was found that the total structural steel quantities could be reduced by nearly 27%. The diagrid provides the added advantage of limiting the number of physical moment connections from approximately 9,600 to 432, resulting in additional savings during fabrication and construction.

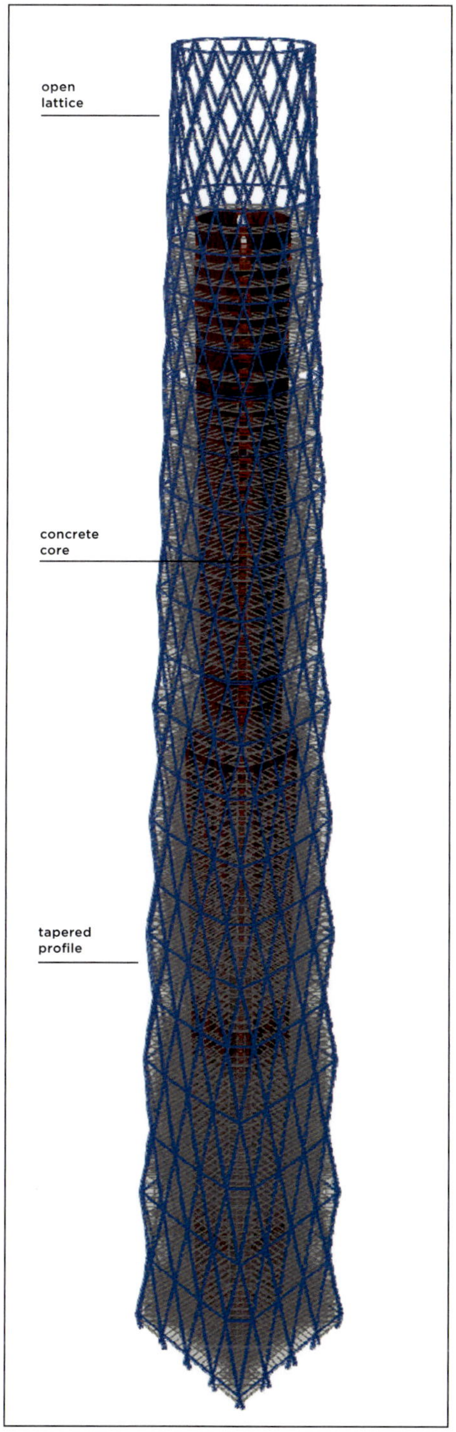

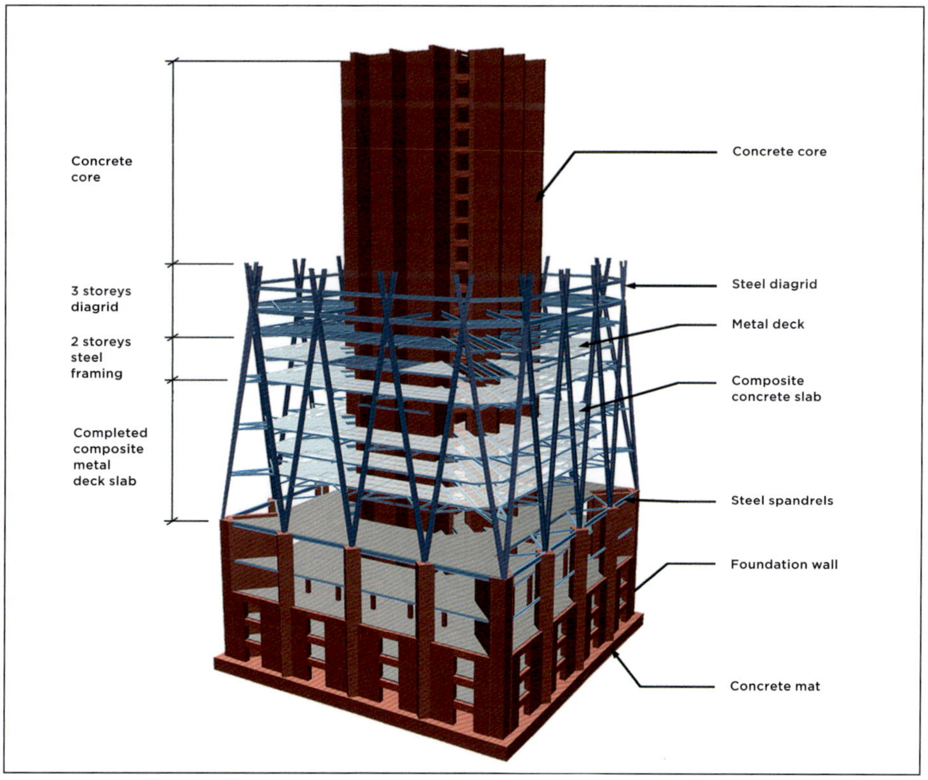

The typical construction sequence as it begins at the base of the tower. The square plan uses the "bird's mouth" approach to the design of the corners, as this eliminates large cantilevered sections of floor plate. As the shape of the tower evolves to its circular plan, this issue is entirely eliminated by the plan geometry.

Nodal configurations were tested with wooden models. The configuration with diagonal member flanges in line with the spandrel (left) was not used. The selected configuration with diagonal members rotated 90° allows for a more compact design.

This project advanced the use of digital modeling as a response to the changing geometries throughout the height of the tower. The team developed custom tools to manage the geometry of the connections between the different parts of the diagrid by creating accurate physical and analytical 3D models of the structure. The team used MicroStation and Bentley structural software to provide the main interface to the custom application used to parametrically describe and manage the node connections. The 3D model supported the creation of 2D and 3D drawings and the provision of accurate digital information to the steel supplier/fabricator.

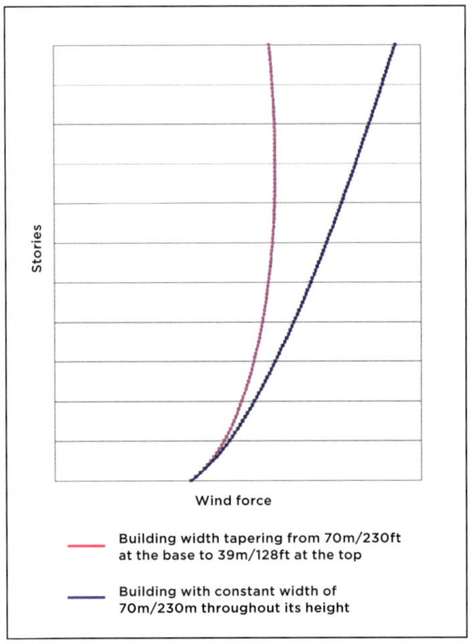

Wind tunnel testing validated the decision to taper the tower as this reduces the effects of the wind load over the height of the tower.

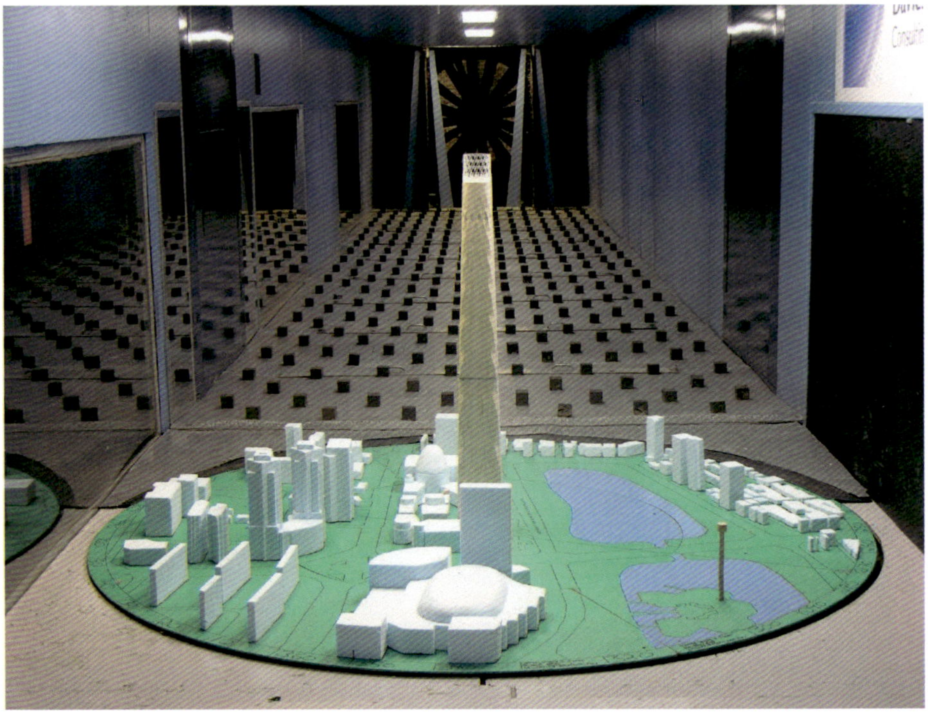

The final design for the tower, undergoing boundary layer wind tunnel testing at the facilities of RWDI in Guelph, ON, Canada.

Exploded view of a typical node assembly, showing the use of custom-welded plates. The detail is expressing an evolved standardization of the type of node connection that is used with H-shaped members, whether they are custom-fabricated or able to use standard hot rolled shapes.

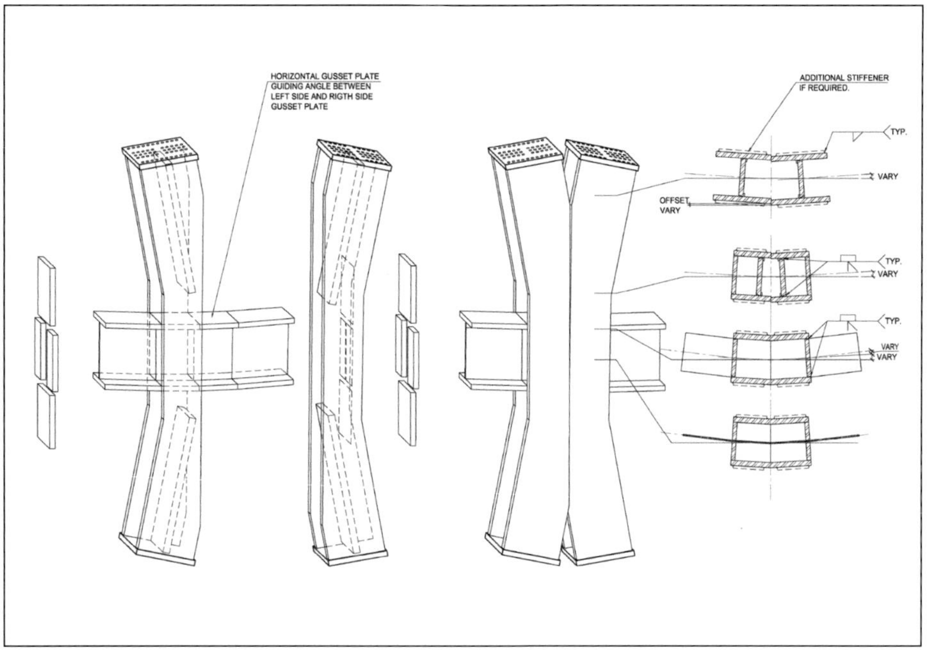

The design of the node system made use of physical and digital models. At each primary level there are either two or three unique nodes, depending on whether the diagrid members are coming together or spreading out at the corners of the building. This leads to a total of 68 unique nodal designs for a condition where six major members come together at a single point. Each unique node requires two angle changes and a rotation for each member coming into it. This allowed all of the geometry in the tower to be based on only 432 nodes establishing all angle changes and rotations in order to simplify the design and construction. The size of any node was limited to 35 tonnes for crane erection. The nodes would be shop-welded from custom-fabricated plate material, with all bolted connections made on site to enhance constructability. With regard to the availability of locally produced steel plate in South Korea, it was desired that a maximum thickness of 80mm/3.15in be specified for any individual plate.

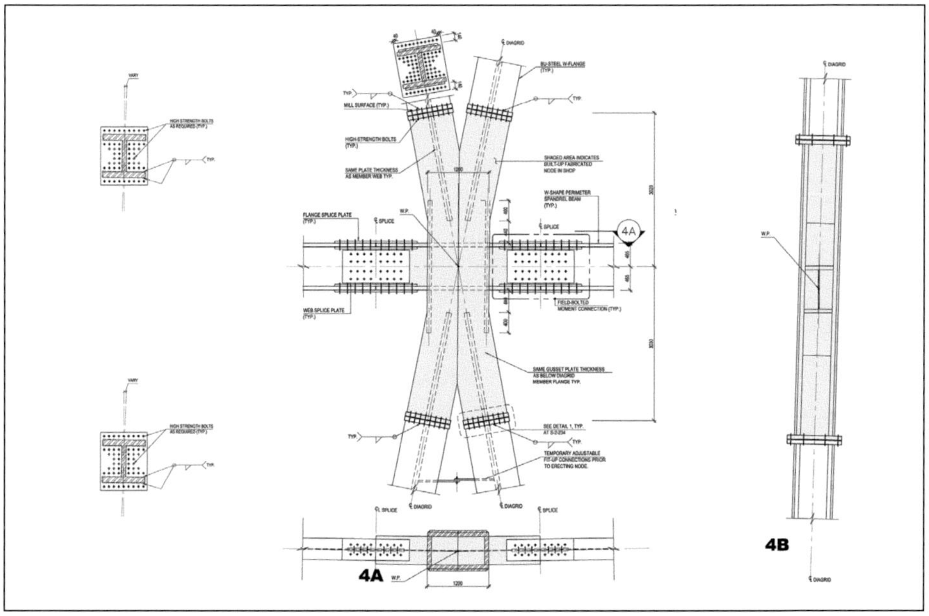

This drawing shows the transformation of the working geometries presented at the top of this page into a highly rationalized nodal design that has considered constructability. The node (shaded in grey) is fitted with milled plates at its upper and lower ends that permit field bolting to the attaching diagonal members. The splice connections to the horizontal ring beams assume a different geometry that uses a site-bolted moment connection. Although the structural elements look like a wide flange configuration, the members would be fabricated from custom-cut plate because of the specific geometries and required plate thicknesses. The system was not designed for architectural exposure due to the fire regulations associated with the extreme height of the tower. Intumescent coatings alone would be insufficient to provide adequate protection.

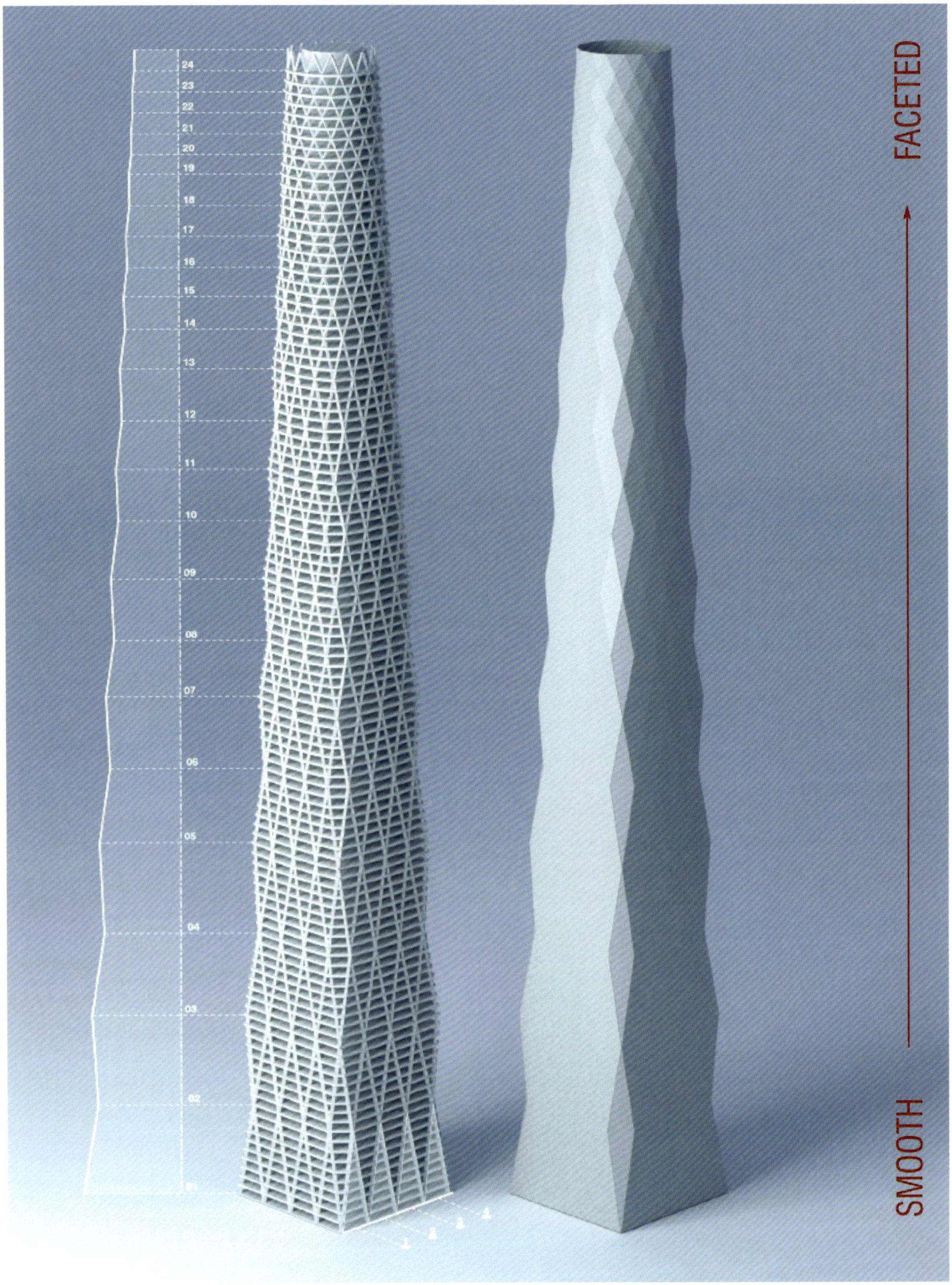

The geometry of the proposed Lotte Super Tower varies throughout the height of the building. The length of a diagonal (a half module) changes from 10 storeys at the base of the tower to two storeys at the top. This reflects a steeper angle of around 78° at the base, where the members must have greater resistance to moment, to a minimum of around 60° at the top of the tower, where shear loads are the greatest.

Each of the components of the structural system was optimized to provide material efficiency and was also geometrically rationalized to increase constructability. This structural diagrid defines the form from bottom to top. According to SOM, "the individual, planar triangles of the diagrid are clad with taut glass surfaces, revealed from their adjacent glass surfaces. Plating the diagrid structure in a geometrically defined, planar surface enhances the transforming geometry from a relatively smooth surface at the square base to a more complex, faceted texture at the crown. This marriage of structure and form is inherent in the building imagery."

The 31-storey Poly International Plaza in Beijing, China, designed by SOM, under construction in 2013.

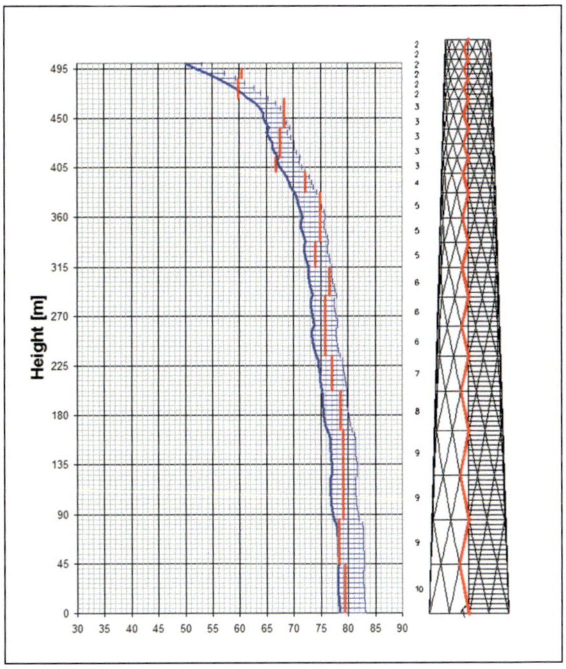

The modules and angles for the diagrid on the Lotte Super Tower vary the angles and module heights over the height of the tower to respond to the changing lateral forces. The angles are steeper at the base and more acute at the top, as the gravity loads and overturning moment forces are more significant at the base while the wind and shear loads are more severe toward the top of this 555m/1,821ft tall tower. The red lines on the diagram represent the rationalized solution that would be used to determine the final angles as averaged from the wider range of theoretical angles (represented by the blue lines) that resulted from loading calculations.

Optimum angle (deg)

— Theoretical Optimum Angle (+5%)

— Rationalized Solution

Although this Lotte Super Tower design for Seoul is not to be constructed, the detailed work is used to inform other SOM-designed towers in preparation or under construction at the time of writing. This would include the 31-storey Poly International Plaza Tower in Beijing, China, which is expected to be completed in 2015. This diagrid tower has an elliptical plan, a four-storey module and relatively shallow angles for the diagrid members – suited to the limited moment forces experienced by a shorter tower type.

A number of important issues have become clear through the detailed design investigations by SOM for the Lotte Super Tower project that can serve to inform subsequent diagrid designs:

- Where shorter diagrid towers do not necessarily require the assistance of a concrete or structural core in resisting lateral loading, taller towers must provide higher resistance to moment toward their base and do require a structural core.
- Toward the top of a tall tower, where shear loads dominate, the structural core can be eliminated and the central area given over to atriums or feature uses.
- Steeper angles should be used on the diagrid members toward the base of the tower, as these are better for moment resistance.
- Shallower angles can be used on the diagrid members toward the top of a tower, as these are better for resisting shear forces.
- Smoother plan shapes and exterior façade conditions are required for taller buildings to reduce vortex shedding issues.
- Advanced structural steel design software and fabrication processes are critical to the success of these complex projects.
- Optimization of the structure and node design does not necessarily require strict uniformity, but rather, careful planning.

Diagrid structures have much to offer as a structural form that is intrinsically connected to a much wider range of design details and decisions. They hold much promise for the creation of engaging future projects.

SELECTED BIBLIOGRAPHIC REFERENCES

GENERAL

Ali, Mir M., and Kyoung Sun Moon. "Structural Developments in Tall Buildings: Current Trends and Future Prospects." *Architectural Science Review*, vol. 51, issue 2, 2007.

Baker, William, Charles Besjak, Mark Sarkisian, Peter Lee, and Chung-Soo Doo. "Proposed Methodology to Determine Seismic Performance Factors for Steel Diagrid Framed Systems." *CTBUH 13th U.S. Japan Workshop*, 2010.

Boake, Terri Meyer. *Understanding Steel Design: An Architectural Design Manual*. Birkhäuser, 2011.

Condit, C.W. *American Building: Materials and Techniques from the First Colonial Settlements to the Present*. University of Chicago Press, 1983.

Mainstone, Rowland J. "Building Tall." http://www.architectureweek.com/2003/0709/building_2-1.html

McCain, Ian. "Diagrid Technology: Structural Efficiency and Increasing Popularity." Available at several sources, among others http://de.scribd.com/doc/129051269/Diagrid-Technology.

Moon, Kyoung Sun. "Design and Construction of Steel Diagrid Structures." *NASCC*, 2009.

Moon, Kyoung Sun. "Optimal Grid Geometry of Diagrid Structures for Tall Buildings." *Architectural Science Review*, 2008.

Moon, Kyoung Sun. "Sustainable Selection of Structural Systems for Tall Buildings." *5th Civil Engineering Conference in the Asian Region and Australasian Structural Engineering Conference*, 2010.

Moon, Kyoung Sun, Jerome J. Connor, and John E. Fernandez. "Diagrid Structural Systems for Tall Buildings: Characteristics and Methodology for Preliminary Design." *The Structural Design of Tall and Special Buildings*, vol. 16, issue 2, June 2007.

National Institute of Standards and Technology. "World Trade Center Disaster Study." http://www.nist.gov/el/disasterstudies/wtc/index.cfm

Nguyen, Binh K., and Hasim Altan. "Strategies to Reduce Lateral Forces on High-rise Buildings that Use Diagrid Structural System." http://bnbarchitects.com

Schneider, Stephen P., Donald R. Kramer, and Douglas L. Sarkkinen. "The Design and Construction of Concrete-Filled Steel Tube Column Frames." *13th World Conference on Earthquake Engineering*, 2004.

Scott, David, David Farnsworth, Matt Jackson, and Matt Clark. "The Effects of Complex Geometry on Tall Towers." *Wiley Interscience*, November 2, 2007. http://onlinelibrary.wiley.com/doi/10.1002/tal.428/pdf

Skyscraper Center. http://skyscrapercenter.com

AL BAHAR TOWERS

Wood, Antony, ed. *Best Tall Buildings 2012: CTBUH International Award Winning Projects*. Routledge, 2012.

ALDAR HEADQUARTERS

Al Dar Headquarters / MZ Architects. http://www.archdaily.com/240524/al-dar-headquarters-mz-architects/

BANK OF CHINA

Blake, Peter. "Scaling New Heights." *Architectural Record*, January 1991.

Tuchman, Janice L., and Robina Gibb. "Stretching the Art of Composite Structures." *Engineering News Record*, October 13, 1988.

Weingardt, Richard G. "Skyscraper Superstar: Leslie Earl Robertson." *Structure Magazine*, June 2007. http://www.structuremag.org/article.aspx?articleID=401

BOW ENCANA TOWER

Charnish, Barry, and Terry McDonnell. "'The Bow': Unique Diagrid Structural System for a Sustainable Tall Building." *CTBUH 8th World Congress*, 2008.

Seica, Michael. "A New Urban Giant: The Bow Tower." *Advantage Steel*, Fall 2010.

CANTON TOWER

Man, Kang, Tony Choi, Vincent Lam, Jackie Yau, William Wong, and Kylie Lam. "Engineering world's tallest freestanding structure." http://www.hkengineer.org.hk/program/home/article.php?aid=6418&volid=141

Tan, Ping, Yanhui Liu, Fu Lin Zhou, and Jun Teng. "Hybrid Mass Dampers for Canton Tower." *CTBUH Journal*, issue I, 2012.

CAPITAL GATE

Schofield, Jeff. "Capital Gate, Abu Dhabi." *CTBUH Journal*, issue II, 2012.

Reina, P. "Wrenches Put in Tower's Works." *Engineering News Record*, July 15, 2009.

CCTV

Arup, ed. *Solutions for a Modern City: Arup in Beijing*. Black Dog Publishing, 2008.

Carroll, Chris, et al. "Case Study: CCTV Building – Headquarters & Cultural Center." *CTBUH Journal*, issue III, 2008.

Carroll, Chris, Paul Cross, Xiaonian Duan, Craig Gibbons, Goman Ho, Michael Kwok, Richard Lawson, Alexis Lee, Andrew Luong, Rory McGowan, and Chas Pope. "CCTV Headquarters, Beijing, China: Structural engineering design and approvals." *The Arup Journal*, 2/2005.

Green, Dane, Bob Lau, Iaian Lyall, David Pritchard, John Pullen, and Clodagh Ryan. "CCTV Headquarters, Beijing, China: Services Engineering Design." *The Arup Journal*, 3/2005.

DOHA TOWER

Wood, Antony, ed. *Best Tall Buildings 2012: CTBUH International Award Winning Projects*. Routledge, 2012.

GUANGZOU INTERNATIONAL FINANCE CENTER

Wood, Antony, ed. *Best Tall Buildings 2011: CTBUH International Award Winning Projects*. Routledge, 2012.

Wilkinson, Chris. "Guangzhou Finance Center: An Elegant Simplicity of Form." *CTBUH 9th World Congress*, 2012.

JOHN HANCOCK CENTER

Iyengar, Hal. "Reflections on the Hancock Concept." *CTBUH Journal*, issue I, 2000.

HEARST MAGAZINE TOWER

Gustafson, Kurt. "A New Angle: Conversation with the Steel Fabricator of Hearst Tower." *Modern Steel Construction*, July 2006.

Hart, Sara. "Norman Foster and the Hearst Corporation Complete an 80 Year Vision." *Architectural Record Innovation*, November 2005.

Lebovich, William. "Hearst Tower." http://www.architectureweek.com/2007/0523/design_1-1.html

Rahimian, Ahmad. "Something Old, Something New." *Modern Steel Construction*, April 2007.

Rahimian, Ahmad, and Yoram Eilon. "Hearst Headquarters: Innovation and Heritage in Harmony." *CTBUH 8th World Congress*, 2008.

Rahimian, Ahmad and Yoram Eilon. "New York's Hearst Tower." *Structure Magazine*, February 2006.

Steel Institute of New York. "Hearst Tower." *Metals in Construction*, Spring 2006.

IBM BUILDING

Nordenson, Guy, ed. by Leslie E. Robertson. "A Life in Structural Engineering." *Seven Structural Engineers: The Felix Candela Lectures*. Museum of Modern Art, 2008.

THE LEADENHALL BUILDING

Young, Andy, Nigel Annereau, Andy Butler, and Brian Smith. "The Leadenhall Building, London." *CTBUH Journal*, issue II, 2013.

Annereau, Nigel, Damian Eley, and James Thonger. "The Leadenhall Building." *The Arup Journal*, 2/2012.

In Progress: The Leadenhall Building/Rogers Stirk Harbour + Partners. http://www.archdaily.com/392494/in-progress-the-leadenhall-building-rogers-stirk-harbour-partners/

LONDON CITY HALL

Barker, Don. "Foster's New City Hall." http://www.architectureweek.com/2003/0226/design_1-1.html

Ferguson, Anthony and Tony O'Meagher. "Fire Engineering at the GLA Building." http://www.mace.manchester.ac.uk/project/research/structures/strucfire/CaseStudy/steelComposite/default.htm

LOTTE SUPER TOWER

Baker, William F., Charles M. Besjak, Brian J. McElhatten, and Preetam Biswas. "555m Tall Lotte Super Tower, Seoul, South Korea." *Structures Congress*, 2009.

Park, Kat, and Nicholas Holt. "Parametric Design Process of a Complex Building in Practice Using Programmed Code as Master Model." *International Journal of Architectural Computing*, issue 3, vol. 8, 2010.

O-14

Reiser, J., N. Umemoto, and J. Ocampo. "Case Study: O-14 Folded Exoskeleton." *CTBUH Journal, issue III*, 2010.

PUERTA DE EUROPA

Winstanley, Tim. "AD Classics: Puerta de Europa / Philip Johnson & John Burgee" *ArchDaily*, September 30, 2011. *http://www.archdaily.com/157275*

Nordenson, Guy, ed. by Leslie E. Robertson. "A Life in Structural Engineering." *Seven Structural Engineers: The Felix Candela Lectures*. Museum of Modern Art, 2008.

MANUKAU INSTITUTE OF TECHNOLOGY

Hammond, M. "Manukau Institute of Technology Building Case Study." *Steel Innovations Conference*, 2013.

ONE SHELLEY STREET

Hewitt, John, Alex Edwards. "The Structural Design of One Shelley Street, Sydney." *The Structural Engineer*, vol. 88, issue 20, 2010.

Wilkinson, Clive. "Future Shift: One Shelley Street." *Indesign*, issue 41, July 2010.

SHUKHOV

Mendina, Alfredo. "Vladimir Shukhov, A Beautiful Mind." http://planta1.com/forum/entry.php?35-Vladimir-Shukhov-a-beautiful-mind

English, Elizabeth C. "Vladimir Shukhov and the Invention of Hyperboloid Structures." *ACSE/SEI Structures Congress*, 2005.

English, Elizabeth C. *Origins of Soviet Avant-Garde Rationalist Architecture in the Russian Mystical-Philosophical and Mathematical Intellectual Tradition*. University of Pennsylvania, 2000.

Harding, Luke. "Foster fires up campaign to save Russian radio tower." http://www.guardian.co.uk/world/2010/apr/15/radio-tower-campaign-russia-foster

Beckh, Matthias. *Hyperbolic Frameworks. Šuchov's lattice towers – forerunners of modern lightweight construction*. Detail, 2013.

SWISS RE

Barker, Don. "Swiss Re Tower by Foster and Partners." http://www.architectureweek.com/2005/0504/design_1-1.html

Foster and Partners. "Modeling the Swiss Re Tower." http://www.architectureweek.com/2005/0504/tools_1-1.html

Ferguson, Anthony and Barbara Lane. "Swiss Re – structural fire engineering for the newest tall building in London." http://www.mace.manchester.ac.uk/project/research/structures/strucfire/CaseStudy/steelComposite/swiss.htm

Munro, Dominic. "Swiss Re's Building London." *Stålbyggnad*, no. 3, 2004.

TORNADO TOWER

Wood, Antony, ed. *Best Tall Buildings 2009: CTBUH International Award Winning Projects*. Routledge, 2010.

ZHONGGUO ZUN

Krummick, Stefan. "Urban Models for the 21st Century." *CTBUH 9th World Congress*, 2012.

NOTES

CHAPTER 2, EVOLUTION OF DIAGRID FRAMING SYSTEMS

1 www.guardian.co.uk/world/2010/apr/15/radio-tower-campaign-russia-foster

2 Whipple, Squire. *A Work on Bridge Building …* H. H. Curtiss, 1847.

3 English, Elizabeth C., thesis, University of Pennsylvania.

4 www.guardian.co.uk/world/2010/apr/15/radio-tower-campaign-russia-foster

5 Condit, Carl W. *American Building: Materials and Techniques from the First Colonial Settlements to the Present*. University of Chicago Press, 1983, p. 187.

6 Condit, Carl W., p. 198. See also www.ctbuh.org/TallBuildings/FeaturedTallBuildings/JohnHancockCenterChicago/tabid/1959/language/en-US/Default.aspx

7 www.structuremag.org/article.aspx?articleID=401

8 www.architectureweek.com/2003/0709/building_2-2.html

CHAPTER 3, PRINCIPLES OF THE CONTEMPORARY DIAGRID STRUCTURE

1 Moon, Kyoung Sun. "Optimal Grid Geometry of Diagrid Structures for Tall Buildings." *Architectural Science Review*, 2008.

2 Moon, Kyoung Sun, Jerome J. Connor and John E. Fernandez. "Diagrid Structural Systems for Tall Buildings: Characteristics and Methodology for Preliminary Design." *The Structural Design of Tall and Special Buildings*, vol. 16, issue 2, June 2007.

3 inhabitat.com/prefab-sky-city-worlds-tallest-tower-to-break-ground-in-june/ and inhabitat.com/worlds-tallest-skyscraper-to-be-built-in-210-days-instead-of-90-as-originally-planned/

4 www.mace.manchester.ac.uk/project/research/structures/strucfire/CaseStudy/steel-Composite/default.htm

5 Dominic Munro of Arup in an article written for the Norwegian Steel Journal *Stålbyggnad*, no. 3, 2004. See also www.epab.bme.hu/oktatas/2009-2010-2/v-CA-B-Ms/FreeForm/Examples/SwissRe.pdf

6 *ibid.*

7 See www.mace.manchester.ac.uk/project/research/structures/strucfire/CaseStudy/steelComposite/swiss.htm

CHAPTER 4, TECHNICAL REQUIREMENTS

1 Ping,, Tan, Yanhui Liu, Fu Lin Zhou, and Jun Teng. "Hybrid Mass Dampers for Canton Tower." *CTBUH Journal*, issue I, 2012.

CHAPTER 7, CORE DESIGN

1 NIST World Trade Center Disaster Study: www.nist.gov/el/disasterstudies/wtc/

2 Dominic Munro of Arup in an article written for the Norwegian Steel Journal *Stålbyggnad*, no. 3, 2004. See also www.epab.bme.hu/oktatas/2009-2010-2/v-CA-B-Ms/FreeForm/Examples/SwissRe.pdf

3 Jeff Schofield, "Case Study: Capital Gate, Abu Dhabi", *CTBUH Journal* 2012, Issue II.

4 Chris Wilkinson in a presentation given at the CTBUH Tall Building Conference in Shanghai, China, in 2012.

CHAPTER 8, CONSTRUCTABILITY

1 Carroll, Chris, Craig Gibbons, Goman Ho, Michael Kwok, Richard Lawson, Alexis Lee, Ronald Li, Andrew Luong, Rory McGowan and Chas Pope. "CCTV Headquarters, Beijing, China. Building the Structure." *The Arup Journal*, 2/2008.

ILLUSTRATION CREDITS

Unless noted otherwise, photographs have been taken by the author.

Cover: The Leadenhall Building isometric, Arup

6 The Leadenhall Building isometric, Arup
10 Tornado Tower, Alexey Sergeev
11 Doha Tower, Ateliers Jean Nouvel and CSCEC
11 Lotte Super Tower, SOM
11 Zhongguo Zun, TFP Architects
15 Bow Encana diagram, Walters Inc.
16 ROM diagram, Walters Inc.
18 Shukhov tower, Sergei Arssenev
23 Tower type diagrams, Vincent Hui
24 Bank of China construction, Leslie E. Robertson Associates
25 IBM Building construction, Leslie E. Robertson Associates
26 IBM Building construction, Leslie E. Robertson Associates
31 Module terms, Vincent Hui
34 London City Hall diagram, Arup
35 London City Hall construction, Arup
35 London City Hall section, Foster + Partners
39 Swiss Re construction, Arup
39 Swiss Re floor plan, Foster + Partners
39 Swiss Re diagram, Foster + Partners
40 Swiss Re lift, Nigel Young, Foster + Partners
41 Swiss Re screenshots (2), Arup
41 Swiss Re construction, Nigel Young, Foster+ Partners
44 Hearst Xsteel node, WSP Cantor Seinuk
45 Hearst Xsteel screenshots (2), WSP Cantor Seinuk
45 Hearst wood node model, WSP Cantor Seinuk
47 Hearst node lift, Nigel Young, Foster + Partners
48 Canton Tower seismic model, Arup
51 Swiss Re wind diagrams (2), Foster + Partners
52 The Leadenhall Building wind testing (3), RWDI
53 Canton Tower wind test, Arup
55 Canton Tower shake table, Arup
55 Guangzhou IFC model, Arup
56 Capital Gate construction, Miroslav Munka
57 Canton Tower construction, Arup
64 Diagrams (2), Vincent Hui
65 Lotte Super Tower diagram, SOM
66 Swiss Re construction, Arup
67 Aldar Headquarters construction (2), William Hare
67 Guangzhou IFC construction, Arup
68 Hearst screenshot, WSP Cantor Seinuk
68 One Shelley Street screenshot, Brookfield Multiplex Australasia
71 Manukau Institute of Technology, Warren and Mahoney Architects
71 Tornado Tower, Alexey Sergeev
72 Aldar Headquarters model and image, William Hare
72 Doha and Tornado Towers, Alexey Sergeev
74 CCTV construction, Arup
75 Canadian Museum for Human Rights screenshot, Walters Inc.
75 Canadian Museum for Human Rights detail drawing, Walters Inc.
76 Capital Gate node digital screenshot, Jeff Schofield
78 Hearst node erection, Nigel Young, Foster + Partners
78 Swiss Re node, Arup
79 Hearst diagram, WSP Cantor Seinuk
82 One Shelley Street construction, Brookfield Multiplex Australasia
82 Aldar Headquarters node screenshot, William Hare
83 One Shelley Street node, Brookfield Multiplex Australasia
83 Al Bahar Towers nodes, Peter Chipchase
83 Al Bahar Towers construction, William Hare
83 CCTV construction, Arup
88 Capital Gate node digital screenshots (3), Jeff Schofield
89 Capital Gate atrium construction, Miroslav Munka
89 Guangzhou IFC construction (2), Arup
90 Canton Tower construction (3), Arup
91 ArcelorMittal Orbit Tower (3), ArcelorMittal
93 Materials diagram, CTBUH
97 Swiss Re plans, Foster + Partners
98 Hearst plans, Foster+ Partners
99 The Leadenhall Building core isometric, Arup
100 Bow Encana diagram upper, Neb Erakovic, Yolles
100 Bow Encana diagram (lower), Walters Inc.
101 CCTV diagrams (2), Arup
103 Al Bahar Towers construction (lower image), William Hare
104 Aldar Headquarters construction, William Hare
104 Puerta de Europa construction (2), Leslie E. Robertson Associates
105 Capital Gate plan and section, ADNEC
105 Capital Gate construction distant view, Jeff Schofield
105 Capital Gate aerial shot, ADNEC
106 Canton Tower construction (2), Arup
106 Canton Tower plan, Arup
107 Guangzhou IFC construction, Arup
107 Guangzhou IFC plans, Wilkinson Eyre Architects
108 Lotte Super Tower interior, SOM
109 Lotte Super Tower diagram, SOM
110 Capital Gate construction, Miroslav Munka
112 One Shelley Street construction (4), Brookfield Multiplex Australasia
113 Guangzhou IFC construction, Arup
116 Capital Gate construction, Miroslav Munka
116 Guangzhou IFC construction, Arup
119 ArcelorMittal Orbit Tower, ArcelorMittal
120 One Shelley Street construction, Brookfield Multiplex Australasia
121 CCTV construction (upper and lower), Arup
125 Aldar Headquarters construction, William Hare
126 Capital Gate glazing mock-up, Jeff Schofield
128 Tornado Tower (left), Alexey Sergeev
128 Doha and Tornado Towers, Patrick Gregerson
131 Capital Gate abseiling under splash, Jeff Schofield
134 One Shelley Street construction (3), Brookfield Multiplex Australasia
137 Canton Tower construction (upper and lower), Arup
138 O-14 all images, RUR Architecture PC
139 O-14 all images, RUR Architecture PC
140 Bow Encana interiors (2), Simon McKenzie
137 The Leadenhall Building interior connection, Arup
142 Doha Tower all images, Ateliers Jean Nouvel and CSCEC
144 Capital Gate view under splash, ADNEC
145 Capital Gate double façade diagrams (2), Jeff Schofield
145 Capital Gate rendering, ADNEC
148 The Leadenhall Building street view, Paul Raftery
149 The Leadenhall Building plans, Rogers Stirk Harbour + Partners
150 The Leadenhall Building node rendering, Arup
151 The Leadenhall Building diagrams (3), Arup
153 The Leadenhall Building overall construction shot/aerial, Dan Lowe
155 The Leadenhall Building distant view, John Safa
156 Capital Gate overall, Jeff Schofield
157 Capital Gate screenshots (2), Miroslav Munka
158 Capital Gate digital model, Jeff Schofield
159 Capital Gate base node, Miroslav Munka
159 Capital Gate node with covers, Miroslav Munka
159 Capital Gate lobby construction, Jeff Schofield
160 Capital Gate 18th floor construction, Jeff Schofield
160 Capital Gate diagrid connecting to core (2), Miroslav Munka
160 Capital Gate atrium construction, Miroslav Munka
161 Capital Gate outrigger truss, Miroslav Munka
161 Capital Gate crane shot, ADNEC
161 Capital Gate node lifting, Miroslav Munka
163 Guangzhou IFC diagrams (2), Wilkinson Eyre Architects
164 Guangzhou IFC digital models (2), Wilkinson Eyre Architects
166 Doha Tower, Ateliers Jean Nouvel and CSCEC
167 Doha Tower plan, Ateliers Jean Nouvel
167 Doha Tower interior, Ateliers Jean Nouvel and CSCEC
168 Doha Tower section and diagrams (3), Ateliers Jean Nouvel
169 Doha Tower diagrams (2), Ateliers Jean Nouvel
170 Zhongguo Zun rendering, TFP Architects
171 Zhongguo Zun renderings (3), TFP Architects
172 Lotte Super Tower renderings (2), SOM
173 Lotte Super Tower diagrams (2), SOM
174 Lotte Super Tower wood models (2), SOM
174 Lotte Super Tower diagram, SOM
174 Lotte Super Tower wind tunnel, SOM
175 Lotte Super Tower node diagrams (2), SOM
176 Lotte Super Tower rendering, SOM
177 Lotte Super Tower diagram, SOM
177 Beijing Poly Building, Brian G. Boake
184 Photo of author, Ian Godfrey

SUBJECT INDEX

INDEX OF BUILDINGS

INDEX OF PERSONS AND FIRMS

ABOUT THE AUTHOR

Terri Meyer Boake BES, B.Arch, M.Arch, LEED AP

Terri Meyer Boake is a Full Professor at the School of Architecture at the University of Waterloo, Cambridge, Ontario, Canada. She is the Past President of the Society of Building Science Educators – an international group of academics that teach in the area of sustainable design. She is Past President of the Building Technology Educators' Society, a group dedicated to vitalizing teaching in the areas of structures and construction.

She has been teaching in building construction and structures since the early 1980s and launched an effort to expand the environmental core curriculum at the School of Architecture at Waterloo during the mid-1990s. She works with the Canadian Institute of Steel Construction and the Steel Structures Education Foundation of Canada to develop interactive multimedia case studies to promote the teaching of steel construction in schools of architecture across Canada. She worked with the CISC Architecturally Exposed Structural Steel task force and has published the *CISC Guide for Specifying Architecturally Exposed Structural Steel* in 2011. She worked to adapt the CISC AESS Guide for the Australasia market and has toured Australia and New Zealand to promote the system. She lectures extensively on Architecturally Exposed Structural Steel. Much of this research was used to inform her recent book *Understanding Steel Design: An Architectural Design Manual*. This theme of AESS will be further developed in her next publication *Architecturally Exposed Structural Steel*, due for publication in 2015.

Terri Meyer Boake holds a professional degree in Architecture from the University of Waterloo and a post-professional degree in Architecture from the University of Toronto. She is an LEED Accredited Professional in sustainable building. Actively engaged in researching carbon-neutral building design, with an emphasis on architectural and passive solutions over a reliance on system and energy sources, a current research project involves the creation of resource materials to assist educators and practitioners in the better understanding of how to create carbon-neutral buildings.

She is an avid photographer and has spent significant time traveling to document and bring noteworthy buildings to the attention of the community. Noting a deficiency and lack of critical construction process information available in current publications, her goal has been to document and present buildings with a "construction eye".